ANTARCTICA

BOOKS BY GABRIELLE WALKER

Antarctica

Snowball Earth

An Ocean of Air

The Hot Topic
(with Sir David King)

ANTARCTICA

An Intimate Portrait of a Mysterious Continent

GABRIELLE WALKER

HOUGHTON MIFFLIN HARCOURT
BOSTON NEW YORK
2013

First U.S. edition

www.hmhbooks.com

First published in Great Britain by Bloomsbury Publishing, 2012

Library of Congress Cataloging-in-Publication Data is available.
ISBN 978-0-15-101520-7

Printed in the United States of America
DOC 10 9 8 7 6 5 4 3 2 1

For Fred and David,
my bookends

All lose, whole find

—e. e. cummings

CONTENTS

Map of Antarctica x

Introduction xiii

Prologue xix

PART 1: EAST ANTARCTIC COAST – ALIEN WORLD

1. Welcome to Mactown 3
2. The March of the Penguins 33
3. Mars on Earth 89

PART 2: THE HIGH PLATEAU – TURNING POINT

4. The South Pole 141
5. Concordia 213

PART 3: WEST ANTARCTICA – HOME TRUTHS

6. A Human Touch 259
7. Into the West 309

Timeline 351

Glossary 357

Notes 361

Suggestions for Further Reading 373

Acknowledgements 375

Index 379

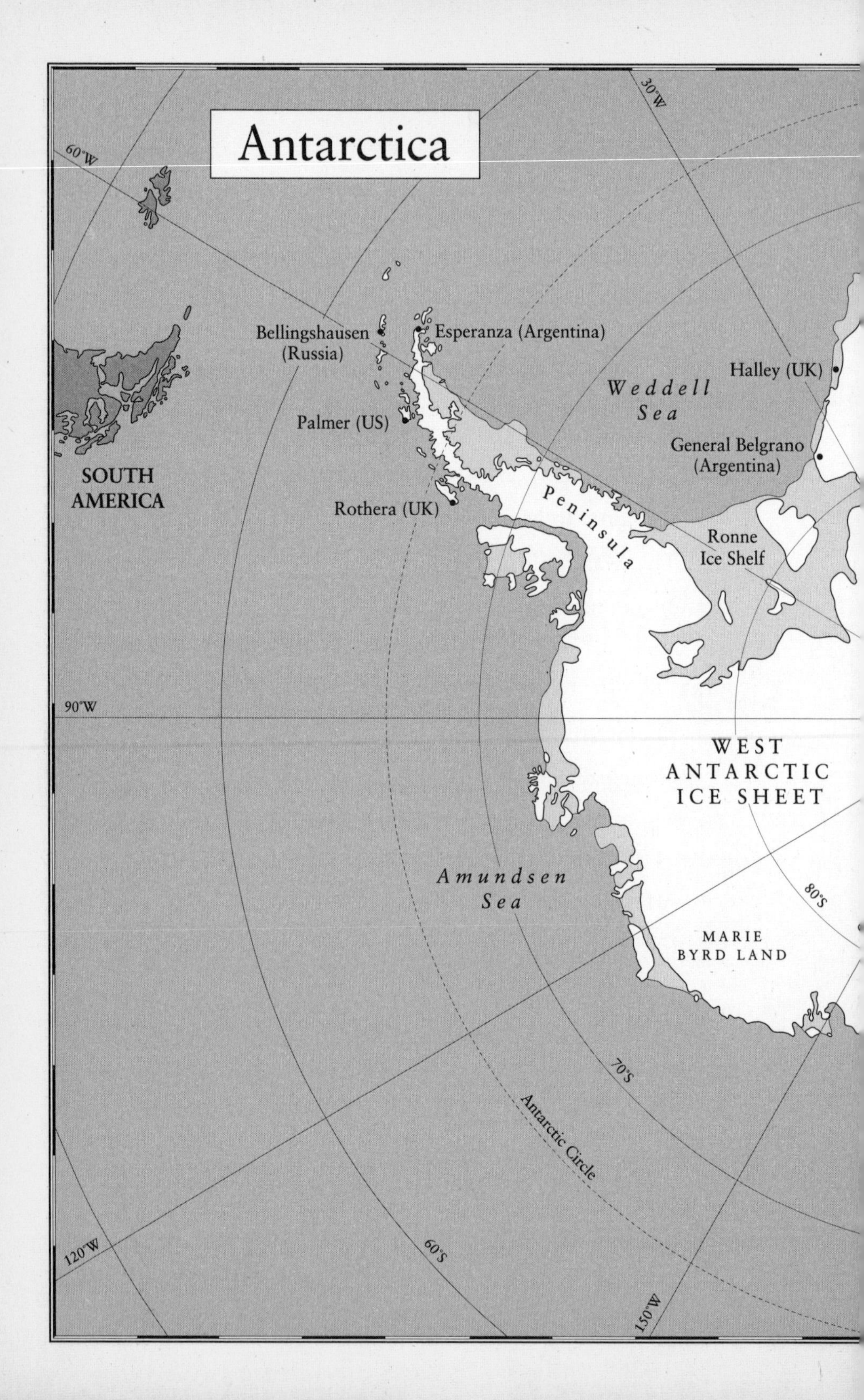

Antarctica
60°W
30°W
Bellingshausen (Russia)
Esperanza (Argentina)
Palmer (US)
Rothera (UK)
SOUTH AMERICA
Weddell Sea
Halley (UK)
General Belgrano (Argentina)
Peninsula
Ronne Ice Shelf
90°W
WEST ANTARCTIC ICE SHEET
Amundsen Sea
80°S
MARIE BYRD LAND
70°S
Antarctic Circle
120°W
60°S
150°W

0°
To Africa
30°E
Sanae
(South Africa)
Troll
(Norway)
Maitri
(India)
Georg von
Neumayer
(Germany)
Syowa (Japan)
60°E
Mawson (Australia)
Zhongshan (China)
Davis (Australia)
EAST
ANTARCTIC
ICE SHEET
South Pole
Amundsen-Scott
(US)
90°E
Mirny (Russia)
Vostok
(Russia)
Ross
Ice Shelf
Concordia
(France–Italy)
Casey (Australia)
Scott Base (NZ)
McMurdo (US)
Ross
Island
Ross
Sea
Mario Zuchelli
(Italy)
120°E
Dumont d'Urville
(France)
To Australia
To New Zealand
180°
150°E

INTRODUCTION

Antarctica is like nowhere else on Earth. While there are other wild places or ones that seem extreme, this is the only continent in the world where people have never permanently lived. In the interior of the continent there is nothing to make a living from – no food, no shelter, no clothing, no fuel, no liquid water. Nothing but ice.

People have long suspected there may be some kind of land at the bottom of the world. The Greeks believed in Antarctica saying, with the peculiar logic of philosophy, that there must be a far southern continent to balance out the land in the north. Poets and novelists dreamed up new races of humans inhabiting tropical southern lands, or a hole at the South Pole that gave access to a hollow Earth beneath.

They were free to dream. The great sailing expeditions of discovery, which showed European powers the new worlds of the west and the ancient ones of the east, were always forced to turn back if they travelled too far south; they were blocked by the great ring of impenetrable pack ice that circles the southern seas.

The first sighting of the continent's outermost islands in 1819 did little to stop the speculation of what might lie beyond, and the first serious attempts to penetrate its interior took place barely a hundred years ago, in the heroic age of exploration by Scott, Amundsen, Shackleton and the rest.

Even now, although this land is bigger than Europe or the continental US, it has only forty-nine temporary bases, most of them on the relatively accessible coast.[1] In summers there are perhaps three thousand scientists on the ice, plus another 30,000 tourists who come in on short visits, usually by ship to the western Peninsula. In winters, there can be just a thousand people on the entire continent.

The scale of the place is hard to grasp. You see a mountain or an island that seems a few hours' walk away and decide to wander over and explore; five days later you're still walking. The early explorers did this a lot. The problem is not just the size of the features – glaciers that make Alaska look small, mountains that dwarf the Alps – but also the absence of anything against which to judge them. There are no trees, or indeed plants of any kind; no land animals; nothing but glaciers, snowfields and sepia-toned rocks.

In spite of its size, Antarctica officially belongs to nobody. An international treaty, signed now by the forty-nine countries with a declared interest, forbids commercial exploitation and dedicates the entire place to 'peace and science'. Thus, the continent is a science playground. Dozens of countries have gained themselves a placeholder for any future exploitation by building bases whose presence is justified by the noble pursuit of science. But whatever the true reasons that governments pump money into Antarctic science, the results extend far beyond the continent itself. Discoveries made there have dramatically changed the way we see our world.

For these reasons and many more I have been fascinated by Antarctica for more than two decades. I have visited five times, mainly as a guest of the huge American programme, run by the

US Government's National Science Foundation, through whom I spent several stints at the South Pole, stayed for four months at McMurdo – the main American base on the coast and the unofficial capital of Antarctica – and visited many of the US field camps scattered around the continent. I've also been a guest of the Italian, French, British and New Zealand governments. I've sailed to Antarctica at various times on a tourist ship, a British Royal Navy icebreaker and a science research vessel. I've driven on the ice in tractors, snow dozers, skidoos and strange tracked vehicles with triangular wheels, and flown over it in helicopters, Hercules transport planes and small ski-equipped Twin Otters.

And in all these experiences I have encountered some astonishing stories. Antarctica has much, much more than just ice and penguins. It is like walking on Mars; it is a unique window into space; it has valleys that time has forgotten; mysterious hidden lakes; under-ice waterfalls that flow uphill; and archives of our planet's history that are unrivalled anywhere else on Earth. It is also a place of romance, adventure, humour and terrible cost. Since there is no prior culture or indigenous population, modern humans can write themselves afresh. For the people who go there, Antarctica is a carte blanche.

Even its apparent barrenness is a large part of its power. People are drawn to Antarctica precisely because so much has been stripped away. The support staff I met there told me that they had come not to find themselves so much as to lose the outside world. The continent lacks most of the normal ways that we interact in human societies. There is no need for money; everyone wears the same clothes and has the same kind of lodging – whether a tent, a hut, a dorm room or, in the bigger bases, an ensuite room that wouldn't be out of place in a Travelodge; you eat the same food as everyone else; you forget about the existence of mobile phones, bank accounts, driving licences, keys, even children. (Almost none of the bases will allow anyone under the age of eighteen.) And with this simplicity of life comes a clarity that's intoxicating.

That doesn't just apply to your time on the ice. A sojourn in Antarctica brings with it a new way of seeing back in the real world. Christchurch, in New Zealand, is the main point of return for the American mega-base, McMurdo. The locals are used to the oddities of Antarcticans arriving after long months on the ice. Nobody is surprised if, while checking into your hotel, you ask for a glass of fresh milk along with your room key (there are no cows on the continent), or if you wander out of a restaurant forgetting to pay. And in the botanical gardens at the end of the season you can often find people sitting for hours, staring in wonder, as if they were seeing flowers for the first time.

With this book I have attempted to weave together all the different aspects of Antarctica in a way that has never been done before: what it feels like to be there; why people of all kinds are drawn to it; Antarctica as place of science, political football, holder of secrets about the Earth's past, and ice crystal ball that will ultimately predict all of our futures. It is only when you see all those different aspects and how they interconnect that you can begin to understand this extraordinary place.

I have tried, in short, to write a natural history of the only continent on Earth that has virtually no human history.

Antarctica is made up of two giant ice sheets. Part One of the book is based around coastal stations on the East Antarctic ice sheet, the larger of the two. This is home to a bleakly beautiful frozen lake district, which is so like the Red Planet that it has been dubbed 'Mars on Earth'. It's also here that you can meet the 'aliens' of Antarctica, creatures that live on the coast there year round and have been forced into bizarre adaptations to cope with the extremes. There are fish with antifreeze in their blood, seals that live out the winter swimming non-stop beneath the sea ice, snow petrels that look angelic on the wing but are spitting maniacs close up, and penguins that put themselves through extremes of starvation and privation to rear each new generation.

For Part Two we move to the high plateau in the interior of the eastern ice sheet. This is where the astronomy happens, giant telescopes high on the summit of the ice sheet that see through windows in the cold, dry sky to parts of the Universe that other telescopes can't reach. This is also where we see how humans pass winters trapped on their bases, as isolated as if they were on a space station.

The fulcrum of the book comes as I describe another treasure found in the east: the extraordinary archive of the Earth's climate history, buried as bubbles of ancient air under three kilometres of ice. While scientists working on the rest of the world were quibbling, Antarctica told us beyond any doubt that our burning of oil, coal and gas has significantly changed our atmosphere, taking it into unnatural and potentially very dangerous territory.

Part Three then focuses on the west of the continent: the West Antarctic Ice Sheet and the peninsula tail pointing to South America. The Peninsula is warming up more rapidly than almost anywhere else on Earth. And the West Antarctic Ice Sheet is the vulnerable one, based on slippery wet rocks that could send it sliding into the sea. Though it is the smaller of the two ice sheets it still contains enough ice to raise sea levels around the world by three and a half metres.[2] If the West Antarctic Ice Sheet melted completely, or even in part, Antarctica would no longer be a remote curiosity. Its ice would fill the oceans, rearing up to flood London, Florida, Shanghai and the hundreds of millions of people who make their livings in places that now seem perilously close to the sea.

The underlying theme of the book is the classic 'hero' story, in which the narrator travels to the end of the Earth, to the strangest, most distant lands, only to find a mirror, the girl next door, the key to life back home. But there is also a deeper message, for which Antarctica is the living metaphor. The most experienced Antarcticans talk not about conquering the continent but about surrendering to it. No matter how powerful you believe yourself to be – how good your technology, how rich your invention – Antarctica is always bigger. And if we humans look honestly into this ice mirror, and see how small we are, we may learn a humility that is the first step towards wisdom.

PROLOGUE

The walls of the crevasse looked grey in the streaky light of Steve Dunbar's headlamp. It was dark and cold and the ice was sheer. The world above him had become cone-shaped, the tapering sides leading down from a distant hole, the size of a manhole, through which daylight was feebly filtering.

Before climbing into a crevasse like this Steve would normally have broken open more of the snow bridge that had masked it, to widen the hole and let in a little more light. But this time he couldn't afford to send a cascade of snow downwards. Somewhere below him in this infernal crack was a human being, who had been down there for thirty hours or more, in temperatures of -31°F. Steve knew what he was likely to find. But still, he had to try.

Word had come yesterday evening, and as soon as Steve's pager had gone off, he'd known it meant trouble. He was head of the Search and Rescue (SAR) team at McMurdo Station. According to his contract, his job was to keep the scientists and support workers on the American research programme safe from harm. According to the unwritten rules of this continent, if anyone anywhere came to grief, the chances were his pager would buzz.

This time it was a Norwegian team. Four of them had been

riding skidoos to the South Pole, hoping to retrieve a tent left there back in 1911 by the great Norwegian hero Roald Amundsen. Amundsen was one of the most famous people ever to set foot on Antarctica, the conqueror of the Pole, the winner of the race to the bottom of the world. This was now 1993 and the tent had been buried by decades of snow, as well as shifted by the moving ice. But the men were confident they could find it, dig it up and take it home in triumph to be displayed next year at the Lillehammer Winter Olympics.

Now, however, they had run into difficulties. A thousand kilometres from their goal, someone had fallen into a crevasse. They had set off a distress beacon, which had rung bells with the Norwegian government, who had called the American government, who had called the US National Science Foundation, who had called the base commander at McMurdo, who had called Steve.

For a nearby emergency, the SAR team could be on the road in about twenty minutes, but the region where the Norwegians were now trapped was about as remote as it was possible to be. While Steve organised a team of seven people and packed up 1,000 lb of gear, the aeroplane coordinators diverted a ski-equipped Hercules from its mission to service a remote science station.

Hercs are heavy planes, far too heavy to take out to a crevassed accident site. This one took the team on the three-and-a-half-hour flight to the Pole, where Steve chose three trusted members – a Navy medic, an American mountaineer and another mountaineer from the New Zealand base near McMurdo – to join him on a smaller Twin Otter plane. They would take some gear, scout out the situation and call in reinforcements as necessary.

By the time they reached the site of the accident, in the Shackleton Mountains, more than a day had passed since the beacon had flared. The pilot spotted a tent and buzzed down low, a hundred feet above the surface, but nobody emerged from the tent. That was a bad sign. There had been radio contact with the Norwegians from the Pole but that had stopped a few hours ago. Through the Twin Otter's window Steve could see countless holes where their skidoos had broken through snow bridges; he could see the tracks where they had hit dunes in the snow and then flown through the air. They must have been going at top speed, vaulting over crevasses, surrounded by danger, holes opening up all around them, scared to death. There were three skidoos parked next to the tent. And about two hundred feet away, one hole had ropes dangling forlornly down into it.

The closest landing site they could find was nearly three miles from the tent. As the Otter taxied after landing, a snow bridge opened up on the left-hand side, leaving a hole that the plane's ski could easily have tumbled into. There were crevasses *everywhere.* Any hopes of bringing in reinforcements now vanished. This was going to be a one-stop mission, to find the casualty, bring him back to the plane and get back out of there.

The team was roped up and ready before they even climbed down on to the ice. Steve took the lead, probing every step with a thin pole almost as tall as he was. His arm quickly grew tired from the repeated lifting and thrusting. The snow was like sugar, so full of air that he could hardly tell where snow finished and hazardous crevasse began. In spite of their care the four of them plunged repeatedly through the snow, their ropes holding firm, their legs dangling over invisible chasms.

And the crevasses were unbelievably chaotic. Instead of the usual parallel lines like stretch marks in the snow, these were a crazy paving of zigzags running every which way. That's really dangerous. Normally you can approach crevasses from the side and then step over them, knowing for certain that even if you fall in, the person behind you won't. But if you can't predict their directions, all four of you could be on the same snow bridge over the same crevasse at the same time. And if you all break through into the same crevasse at once, everybody falls. Steve's sense of responsibility grew heavier with every dogged step. At what point did his obligation to protect the people behind him on the rope begin to override his obligation to help the people he'd come to save?

He kept going; they all did. Four hours of slog just to travel three miles. When they were just a few yards away from the tent, two of the Norwegians finally climbed out to greet them. Steve could tell straight away that they were shot to pieces emotionally. Inside the tent, one of their companions had cracked ribs and concussion. He'd been the first to fall. His skidoo had broken a hole big enough to plunge into and he had gone with it. Luckily for him he had smashed into a ledge in the crevasse and stuck there, unconscious, while his skidoo crashed on down into the abyss. When he came to, he had managed to climb out using a chest harness and ropes that his companions had thrown to him. A chest harness with broken ribs? That must have been agony.

It was after this that the others had set up the tent. But then the real disaster hit. The team's second in command, an army officer named Jostein Helgestad, had decided to try to find a safe passage through the crevasses on foot. His companions had seen him disappear into the ice just a stone's throw from the tent. And they had heard nothing from him since.

Somebody had to look, so Steve secured a rope to one of the skidoos and went in. Sixty feet down, the crevasse was so narrow that he couldn't turn his head for fear of knocking off his head-lamp; the danger was now not falling so much as getting wedged in. His legs were splayed, his crampons snagging on the ice walls. He couldn't control his own rope any more; his companions up on the surface were going to have to start lowering him. He yelled up instructions then pivoted vertically so that he was descending head first into the darkness. His headlamp picked out a sleeping bag that had been thrown down by the Norwegian team and had evidently been left untouched. Then he saw his man.

The crack was now barely a foot wide. Jostein was wedged in sideways where he had fallen and where his body heat must have melted him further into the ice. Steve strained to touch him but couldn't quite reach. Instead he probed down with his ice axe, snagged Jostein's arm and gingerly raised it. The arm was frozen solid.

No hope, then, of even retrieving the body, but there were still three people to save. Back on the surface, Steve gave more instructions. The Twin Otter was neither big enough nor fuelled enough to take a heavy load. The only way they could all get out was to abandon everything. They left tent, skidoos, clothes, harnesses, ropes, everything except the gear they needed to get to the plane. Steve found himself explaining the principles of roped glacier travel to a man who was still seeing double from concussion and two others who were dazed at the disaster that had befallen them.

And then there was the long perilous slog back, a careful check for crevasses along an improvised runway, more gear ditched, more load lightening, and a take-off for which everybody held their collective breath before the plane finally rose into the air over Antarctica's bright white hinterland.

Nearly twenty years later, Jostein Helgestad is still there, his frozen body held fast by a continent that punished his boldness without hesitation or particular interest. The truth is that Antarctica has little time for humans. We have managed to colonise most of our planet, to get by in apparently hostile deserts, forests and mountains. Even at the North Polar ice cap, which is a frozen ocean surrounded by continents, the sea ice is just a thin skin and the animals that swim beneath have provided humans with food and fuel and clothing for thousands of years. But Antarctica is different. It is a vast, isolated stretch of rock, almost completely buried under thousands of feet of ice. This is the only continent on Earth where people have never lived. And until very recently in human history it was as mysterious to us as the Moon.

Even today, the temporary bases that dot the continent are miniature life-support systems, human toeholds on the edge of a vast, alien landscape, for which everything you need to survive has to be brought in from the outside. Yet people still go there in their thousands every year, as scientists, explorers, adventurers and the incurably curious.

But curiosity can also be perilous. And if you do find yourself in trouble, the phone will again be ringing at McMurdo Station, the biggest of all the bases, logistics hub, unofficial capital of Antarctica and gateway to the ice.

1

EAST ANTARCTIC COAST

Alien World

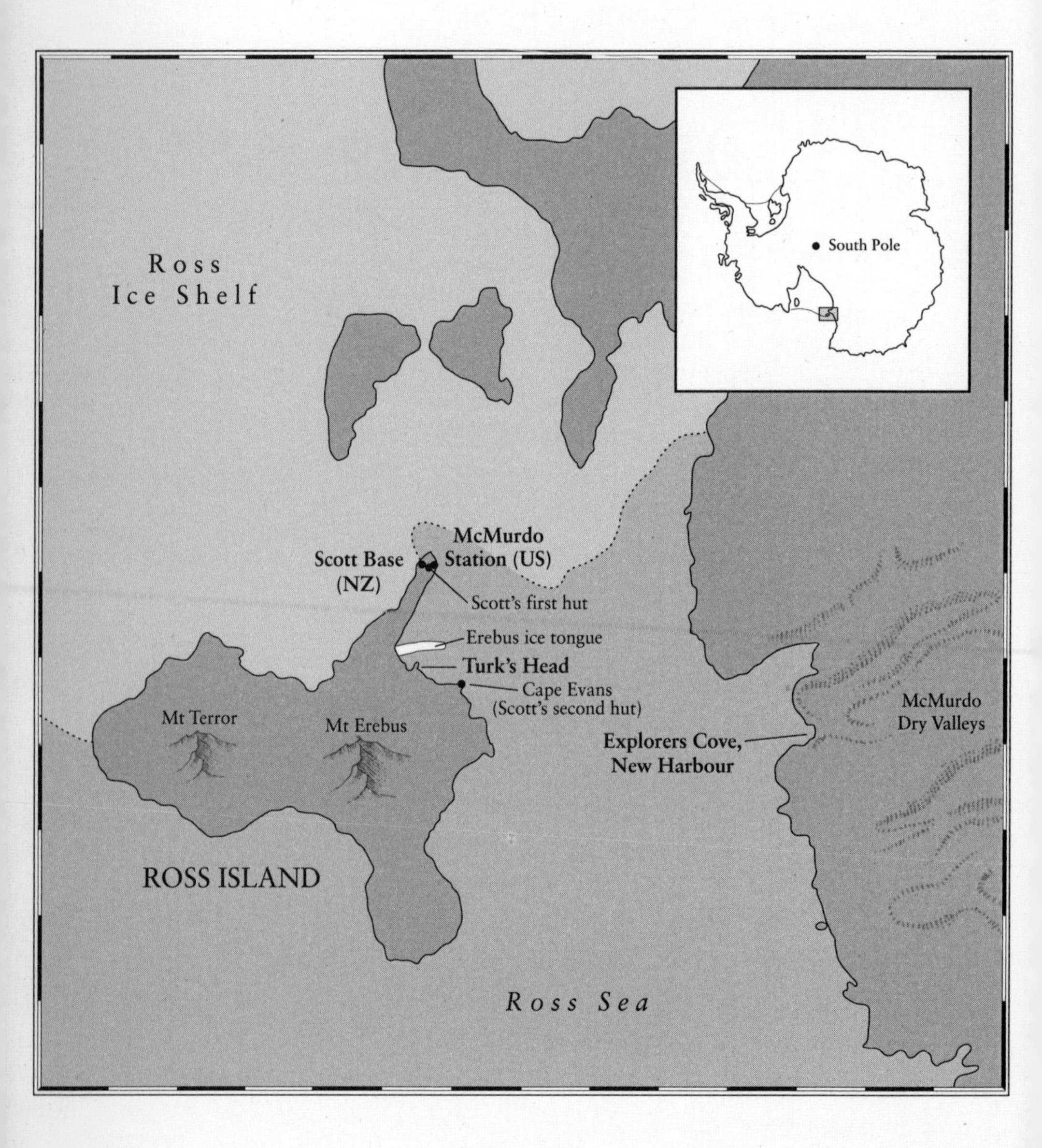
Ross
Ice Shelf
South Pole
McMurdo
Station (US)
Scott Base
(NZ)
Scott's first hut
Erebus ice tongue
Turk's Head
Cape Evans
(Scott's second hut)
Mt Terror
Mt Erebus
McMurdo
Dry Valleys
Explorers Cove,
New Harbour
ROSS ISLAND
Ross Sea

1

WELCOME TO MACTOWN

McMurdo Station lies on a volcanic island, as far south as you can sail from New Zealand before bumping up against Antarctica – which is how the earliest explorers discovered it. These days, however, most people fly there, in big, noisy, military troop transporters, strapped into webbing seats and packed around with cargo.

If you're lucky, you'll get through first time. If you're unlucky, the weather will turn bad just before the plane reaches the 'Point of Safe Return' at which there is still enough fuel to make it home, and you will boomerang back to New Zealand, for another long, uncomfortable try tomorrow. (The far end of the boomerang used to be known as the 'Point of No Return', but was changed for purposes of reassurance.)

Known to its inhabitants as Mactown (or just 'town'), McMurdo is the operational headquarters of an American research programme that reaches out from here to the entire continent. But if this is your first sight of Antarctica, and you're expecting great sweeping vistas of snow and ice, you're likely to be surprised.

Coming in from the sea ice runway, on a massive bus whose wheels are taller than your head,[1] you bump endlessly over invisible obstacles, craning your neck to try to peer through the windows. But they are hopelessly steamed up by the crowds of people around you, who are all quietly overheating in the many regulation layers of clothes they have been obliged to wear in case of breakdown.

And then at last you arrive, and tumble down the steep steps of the bus to see . . . a grubby, ugly mess. McMurdo itself has no ice and little romance. It is more like a mining town, planted squarely on dirt. The buildings are squat and mismatched, with tracked vehicles and heavy plant lumbering along the roads in between, churning up the black volcanic soil and spreading dust and grime. There is nothing to soften that hard industrial edge. You will find no trees or other vegetation here, and nor are there children or non-native animals. All foreign species other than adult humans are banned.

I remember my first few hours at Mactown, but they were also strangely blurred. There was a constant buzz of helicopters overhead; trucks were shifting materials from one building to the next. People were running past, dragging the big orange bags that were issued to everyone back in Christchurch, to carry the regulation red parka and wind pants, and thermals, and water bottle, and a bewildering array of gloves and mitts and scarves for every occasion. Others were heading down to the sea ice on skidoos that roared like motorbikes. And we newbies were trying to fill in the many, many forms, and take in the dizzyingly detailed instructions about where we needed to be, when, and why, and with what.

At one o'clock in the morning when everyone else had finally got off to their allocated dorm rooms to sleep, I stole away in the bright midnight sunshine to the edge of town and climbed up Observation Hill, a local cinder cone shaped like a child's drawing of a volcano.

The path was rocky but clear and after about an hour I reached the summit, marked by a tall wooden cross. This was erected back in 1912 by the colleagues of the doomed Captain Scott, after he lost his life on the way back from the South Pole. It was inscribed with the names of the five men who perished, along with a line from 'Ulysses' by Alfred, Lord Tennyson: 'To search, to seek, to strive and not to yield'.

Scott based his two Antarctic expeditions on Ross Island. The second expedition, the more famous of the two, started from Cape Evans, around the coast from here. But the first was built at 'Hut Point' at the end of the peninsula in front of McMurdo. I could see it now over in the distance, near where the icebreaker ship docks on its annual resupply. With its clean wooden walls and tidy low roof, the hut looked as if it were built yesterday – a reminder both that ice is a great preserver, and that the heroic age of Antarctica wasn't so very long ago.

I sat with my back to the cross and thought about the many dramas that had taken place around that small hut. The messages fixed to the door for returning field parties, which had spoken of disaster at least as often as triumph. The people who had trudged and strained their way over the ice, hoping for bright lights and a warm welcome, and found only darkness.

To the left I could see the white expanse of the great Barrier, a floating glacier the size of France that we now call the Ross Ice Shelf. Its edges form giant cliffs of ice above the ocean (and even bigger stretches of ice below), which prevented the early explorers from sailing farther south than here. Instead, any attempt to reach the South Pole meant slogging for hundreds of miles over its surface, a place that one of Scott's men described as 'a breeding place of wind and drift and darkness'. It was on the Ross Ice Shelf, some ninety miles from here, that Scott and his men finally succumbed to cold and hunger.

About two hundred miles to the east, Roald Amundsen set up his rival camp, not on solid ground like Ross Island, but on the surface of the Barrier itself. Though most of the cliffs are impenetrable, Amundsen had found one inlet, called the Bay of Whales, which gave him a way in.

Scott didn't know he was in a race to the Pole until his men spotted Amundsen's ship, the *Fram*, while Scott and others were laying supply depots elsewhere on the Barrier. The two crews politely shared meals and plans, but the British were soon hurrying back to base with the news that they now had a rival. Scott was deeply shocked. He had thought he had a clean run at glory and although he made as light of it as he could, his men privately recorded in their diaries that he seemed to be sleeping badly, and that the news had obviously hit him hard.

Throughout the winter, each in their own camp, the two teams stocked and tallied and prepared. Amundsen had brought a small, crack team. Everyone knew their role and they spent the dark, cold months refining their equipment. Scott had brought three times as many men, including two who were paying for the privilege, and their activities were more muddled. Practising or trying too hard seemed almost ungentlemanly. Even before they started out for the Pole, the seeds of the coming tragedy were already being sown right here, on Ross Island.

But although Amundsen made the right choices, and ultimately won the race, it is the spirit of Scott's hut that survives in Mactown today in the overwhelming focus on science. Scott's men spent much of their winter giving each other scientific lectures and he himself wrote that, because of the science his men planned to do during their expedition, 'If the Southern journey comes off, nothing, not even priority at the Pole, can prevent the Expedition ranking as one of the most important that ever entered the Polar regions.'[2]

Today, science lies at the heart of the programme. In summer there can be 1,200 people in McMurdo, in winter perhaps 250; and they are all here with one overriding purpose: to support the

US National Science Foundation (NSF)[3] and the scientists that it selects to bring down here.

The Antarctic Treaty, signed by twelve nations in 1959 (and later ratified by a further thirty-seven), bans military or commercial activity, including prospecting. The wildlife is protected, and everything brought in must eventually be taken out. To be allowed to have any significant presence here on the continent, a government must sign the treaty and set up scientific research.

There are some who mutter that the science is just a placeholder, an excuse to plant a flag and maintain a presence just in case Antarctica proves strategically valuable for some other reason. But the selection process for science grantees is intensely competitive in every participating nation. Nobody can come here unless they have proved themselves in many rounds of testing.

And if you ask the support workers why they are prepared to abandon their home life and come here to this ugly town to work six long days a week for months at a time, one person after another will tell you they love the sense that they are doing something that matters, and the chance they have to learn. Science is the lifeblood of the base. It pervades everything. And the talks given by scientists in the galley after dinner tend to be standing room only.

In spite of – or perhaps because of – its isolation, Antarctica turns out to be a fantastic place to do science; over the years it has yielded extraordinary insights into our world. Working in field camps and bases throughout the continent, researchers from many different countries have explored the hostile and the alien, and have found new ways of seeing everything from the Moon and Mars to the heart of the Galaxy and the origins of the Universe. But the farther in you go, the more that home tugs at you; the continent's icy mantle has messages not just about outer space, but about the history of our own human world, and perhaps also its future.

In the process, the ice has revealed to us many of its most extraordinary characteristics. Science is only one of Antarctica's faces, though it's the main one that the world currently sees. But there is also history, politics, natural history, romance and adventure. You might ask which of these represents the true face of the continent. The answer, of course, is all of them.

As well as the scientists and contractors, Mactown also has a steady trickle of VIPs, artists, musicians and writers whom the NSF has invited to provide a new view of the continent. My office mate for the first few days was Yann Arthus-Bertrand, the French photographer who has made his name taking extraordinary photographs of Earth from the air.

When I arrived he had already spent a month photographing the area around McMurdo from helicopters. But, he told me, he also had a side project, a video exploration of what makes people tick. He had taken his standard set of questions around the world, and planned to distil the answers and display them in some sort of installation. On my first day in the office, Yann sat me down, pointed a video camera at me and then reeled off the questions:

'What is your greatest fear? Do you feel you give enough love to the people around you? What could you never forgive? When is the last time you cried and why? Do you have enemies? What is the meaning of life? Are you happy? What does money mean to you? Why is there poverty and why do we tolerate it? What do you think there is after death? Who do you hate and why?'

Afterwards I asked him what he had found so far, and he told me that the residents of McMurdo had thrown up two answers that took him by surprise. First of all, in spite of the many grumbles, an astonishingly high proportion of people here said they were happy. But the money question really surprised him. Elsewhere in the world people usually said that money means power, or security, or status. But to the people of McMurdo, money – apparently – meant freedom.

McMurdo is paid for by the US government and is host mainly to American scientists with their occasional international collaborators. For them it is a staging post, a gateway out into the field. Most stay here for just a few days to pick up their gear and do the obligatory training. There's a two-day snow school to teach you how to pitch tents, light primus stoves and work the bulky high-frequency radios that might be your only way to signal for help if a helicopter or plane crashes or you get trapped outside in a storm; then there are compulsory briefings on the various forms of Antarctic travel. (The helicopter briefing shows the position to adopt if you're heading for a crash. 'It's the classic "kiss your ass goodbye",' the instructor said when I took my course. 'Don't come out of this position until everything stops, or until you hear the pilot say "whew, that was a close one" . . . And even then you might want to give it a few minutes.')

Researchers collect their field equipment from a vast warehouse of tents, sleeping bags, primus stoves and ultra-cold-weather gear, and choose menus from a similarly vast frozen food store. And then they climb aboard the helicopters or ski-equipped planes that will take them to whatever outpost they have chosen to study.

But some science happens right here, on the edge of town. Beneath the sea ice, for instance, where alien creatures, the weird and the wonderful, are willing to go to extreme lengths to make this most hostile of continents their home.

They have spiders the size of dinner plates! Giant slimy worms twice as long as I am tall! Creatures with flailing legs and crushing mandibles that are bigger than my hand! If I'd known Sam Bowser's fondness for science fiction B movies when I first met him, I'd have suspected him of spinning me yarns. But the evidence is there. I've seen the home movies he has made while diving under the sea ice off the coast here, and I have stuck my hands in the freezing waters of the Crary Lab aquarium to pull out some of the bizarre animals he describes.

Sam is a biologist from New York State's Wadsworth Center in Albany, and he has been diving in Antarctica for years.[4] Each season his team sets up camp across the sound from McMurdo, in a place called Explorers Cove. This is not regular ice diving – you can't just saw out a hole, because the sea ice here can be three, four, even five metres thick. Instead they drill a thin column down through the ice, adding extensions to the drill like the brush from an old-fashioned chimney sweep. Then they insert a sausage string of bright red explosives and Boom! There's your dive hole.

The water here is around 28°F, as cold as it's possible for the sea to get; the salt allows it to dip below the normal freezing point of water, and the ice floating above it keeps it from warming. It's physically painful to keep your hands in for the count of ten. And Sam's team dives in these temperatures for up to an hour at a time.

He says the trick is to wear layer upon layer of thermals under a crushed neoprene dry suit. The hands are the hardest to keep warm. Hand-warmers help, and several pairs of gloves, though if you wear the really warm ones – the orange three-fingered monstrosities that turn your hands into lobster claws – it can be hard to handle any equipment you take down there with you.

Your mouth is the only part of your body exposed directly to the water. At first it hurts a bit and then it goes numb and the problem disappears. But when you re-emerge from the dive hole, your lips will be rubbery and useless and if you try to speak in the first few minutes, your words will be comically garbled.

But this discomfort is worth it, Sam says, for the sights you see as you emerge from the tunnel of the dive hole. 'It's incredible. There's this giant ocean below you. It's like walking through a spaceship door and seeing the universe. If you're not going to be a space man, you'd better be an Antarctic diver, because it's the next best thing.'

The water is so clear that you can see for 250, maybe 300 yards in the green half-light. Your head tells you that this is impossible, that distant divers cannot be so far away and still so clearly visible, that they must instead be much closer, hanging nearby in the water like tiny Tinker Bells. Nobody is tethered. You float freely to maximise your flexibility, always deeply mindful of the shaft of light, what Sam calls the 'Jesus beam', that shoots down from the dive hole and shows the way home.

The underside of the ice is sometimes flat, sometimes cathedral-like, with columns of stalactites and feathered frosting. Your air bubbles rise up and collect beneath it to form silver puddles like mercury, surrounded by water.

And below you, the grey sea floor is carpeted with alien creatures. Run your flashlight over them and their colours leap out. There are brittle stars, golden discs that raise themselves up on their five long legs as you approach, and then march away on tiptoe like the Martians from *The War of the Worlds*; feather stars, 40 cm across, that look like a bundle of bottle brushes and swim by waving their protuberances wildly as if they were drunken octopuses; and sun stars, a sort of bright orange starfish with up to forty arms, which in the waters of McMurdo Sound can grow to a metre or more.

For this is the land of the giants. The sea spiders here are more than a thousand times bigger than the ones elsewhere in the world. They stride over the seabed like colossi, a full foot from tip to toe. They are supposed to have eight legs, like their relatives on the land. But some have ten or even twelve. (When the first naturalists came back from Antarctica in the 1820s with drawings of these beasts, their colleagues thought they had accidentally drawn too many appendages.) They look like tall, spiny tarantulas and they are unexpectedly beautiful.

And then there are the less picturesque creatures. The ribbon worms, which are as thick as your thumb and can grow to ten feet long, a writhing, revolting mass of toxic slime, like naked intestines squirming around the sea floor. Ribbon worms occur in most parts of the world's oceans but the ones here are the heftiest of all. They chase limpets. They've even been known to catch fish.[5] Or there's *Glyptonotus antarcticus*, which look like woodlice, or perhaps cockroaches, except that they are bigger than your hand, with a crunchy hard shell and creepily crawling feathered legs that wave wildly when you turn them upside down. Their mandibles reach menacingly from under their leathery segmented shells, and they are said to have inspired the creatures in the *Alien* movies.

Why should these creatures be so much bigger than anywhere else? Although it seems paradoxical, the answer lies in the extreme cold temperatures. Life here is necessarily very slow. Chemical reactions take place at an absurdly sluggish rate, and animals can live very much longer than their warmer cousins. On top of that, colder water can dissolve more oxygen, which is essential for growing big. It's as if the cost of living is cheaper here, in the cold outpost, away from the fast-paced cities, so everyone can afford to have a bigger house.

Still, living in water that lies below the normal freezing point has its own dangers. The spiders and other invertebrates don't mind too much. Their bodily fluids tend to be quite thick, and well protected from infiltration by ice crystals in the water. But fish are another matter. They have to take freezing seawater into their stomachs to get the salt they need, and their thin blood is no barrier to the infusion of cold from the outside. So they have had to develop another, decidedly weirder adaptation than simply growing large: they fill their bodies with antifreeze.

This discovery was made by another McMurdo researcher, Art

Devries, back in the 1960s in several fish, including one called *Dissostichus mawsoni*.[6] You can see these on any dive near McMurdo. They have large blockish heads, jutting jaws and thick lips; they can grow to more than a metre long and they're often called Antarctic cod (although they are actually no relation). Art found that their thin blood is swimming with home-made anti-freeze, not unlike the stuff you would put in your car. It shows up in their stomachs, gut walls, liver, brain, the whole body. This clever strategy only protects them down to about 28°F, though, and if you catch one on a really cold day and accidentally touch it to the ice on your way back up, it will freeze solid immediately – and become a fish popsicle.

But Sam Bowser's favourite Antarctic sea creatures, though giants in their own world, are only just visible on the dives. He spots them by planting a flashlight with a spike into the sea floor so that its beam shines out horizontally and makes them sparkle into sight. They are called foraminifera, or forams for short. Some look like miniature oak trees; others are floppy and filled with droplets of fat, and all have built themselves protective coats of glittering grains of sand.

Sam is fascinated by these odd creatures in part because they are much larger than they have any right to be. Each of them is made up of one single cell. They ought to be microscopic, smaller than the full stop at the end of this sentence. But they are actually several millimetres long, the size of a fingernail, big enough to pick up individually with a pair of tweezers.

They are clever, too, in the design of their protective coats. Sam has cleaned off the sand, put them in a dish of water with some differently sized glass beads, and watched as they extrude sticky tentacles called pseudopodia, literally 'false feet', that pick up the beads and draw them in. The tentacles move using tiny motors of the kind that propel sperm. And they pulse, almost as if they were dancing, as they select a deliberate sequence of small and large beads so that the overall shell hangs together. Each of these creatures is made up of *one single cell.* And yet they are master masons.

Mainly, though, Sam likes the forams because their way of adapting to these extreme conditions is to be extremely, presumptuously ferocious. In the tight-knit food web of Antarctic waters, they punch very far above their weight.

He discovered this one day while doing simple experiments with forams that he had brought back to the Crary Lab. 'Single-celled organisms should be at the bottom of the food web,' he says. 'If they eat anyone else it should be other single-celled organisms, you know, like bacteria, or algae. So we just wondered what other organisms eat these forams. We found some likely things, crustaceans, and left them together in a petri dish.'

The following morning he discovered that things hadn't gone entirely the crustaceans' way. 'We found an even bigger foram, and the shattered remains of the crustaceans. These forams are predators! They rip the flesh out of the much bigger critters that land in their webs. We've done time-lapse movies, it's really gruesome.'[7]

The forams do this using the same sticky pseudopodia that draw in sand grains for their shells. If you are a small crustacean, just passing by, you might land on these, feel irritated and try to wipe yourself off. It doesn't work. Extra tentacles take hold. You start writhing now, but the pseudopodia are like fly paper and the more you struggle the more doomed you are. And then, when you are well and truly trapped, the pseudopodia start looking for parts of your body that they can penetrate. 'They go *everywhere*,' Sam says and then he laughs darkly. 'They start pulling off pieces of flesh. They rend you while you're still alive, and pick you to pieces.'

This is emphatically not the way the food web is supposed to work. Multiple-celled creatures are supposed to eat single-celled creatures, not the other way round. Sam and his group tried other potential predators: juvenile starfish, juvenile urchins. Everything they threw in became stuck and then became lunch.

This may seem ghoulish, but it's also a very effective strategy for surviving winter in Antarctica: capture something much bigger than you are, and fatten yourself up.

Sam and his team are now putting a camera into the water to see if they can watch this happening in the wild. They also want to know if the tree forams are rooted or if they roam over the sea floor. 'I hope they walk,' he says. 'That'll be a fun story.'

Stay in McMurdo a little while and you will start to see beyond the ugliness and the dirt. In summer it rarely snows there, but the occasional dusting does wonders for the place, and it spends its entire winter blanketed in white. Residents are fond of saying that Mactown looks better 'with its clothes on'.

And if you look outwards from town on a clear day, the views are just gorgeous. To the south are snow-draped islands half-smothered by the floating Ross Ice Shelf. To the west, across a frozen sea, lies a line of jagged snowcaps, the tail end of a range of mountains that bisects the continent. It is typical of this odd place that such a glorious highland should rejoice in the prosaic name 'The Royal Society Range', after the learned society that sponsored some early expedition.

The names of Antarctic features can be delightfully eccentric the continent over. There is also an 'Executive Committee Range', which is apparently just as beautiful as the Royal Society's, and 'The Office Girls' are two mountaintops poking out of an ice cliff, which were named in 1970 'to express appreciation for the dedicated support provided to Antarctic programs by home-based personnel'. Other names can be merely descriptive, either of the feature itself – 'Brown Peninsula', 'White Island' –

or ones where the emotions evidently experienced by the discoverers begin to creep in: 'Desolate Island', and 'Cape Disappointment', and – my personal favourite – 'Exasperation Inlet'.

And the town has a unique charm, if you let yourself see it. Though it has a slight military flavour (the base used to be run by the US Navy, and the canteen is still universally called 'the galley'), if anything McMurdo feels more like a university. It has dorms and bars and craft centres and a general sense that the life here is both intense and temporary.

That fits the people, too. Most of the contract workers are unremitting dreamers. They have come seeking adventure and do their best to find it in among the soapsuds or engine grease of Mactown. They work as dishwashers, hairdressers, bar attendants, cleaners, heavy equipment mechanics and locksmiths. It's not hard to find someone with a Ph.D. peeling potatoes in the galley, or a qualified lawyer counting widgets for Supply. And they are all desperate to get out of McMurdo and into the wild white beyond.

These occasions don't arise easily unless you are a scientist. But few people complain about it. McMurdo residents have a saying. 'It's a harsh continent,' they will tell you, if you complain about regulations, or your hours, or if the ice-cream machine breaks down. It's deliberately ironic; to claim to be tough here is to invite scorn.

And yet, McMurdo wasn't nearly as macho as I'd expected. I'd heard the mock slogans: 'Come to Antarctica – where there's a woman behind every tree!' Or how the few women there earned themselves ratings, in which an 'Antarctic 10' was someone who would rate a '5' elsewhere in the world. But, surprisingly, there were plenty of women at Mactown and I felt utterly comfortable in the bars, the coffee shop, or just wandering around town.

It wasn't always this way. For more than a hundred years after men first walked on Antarctica, women were not allowed anywhere near it. The first woman known to set foot on the continent was Caroline Mikkelsen, wife of a Norwegian whaling captain, who briefly went ashore in 1935. The next was not until 1947 when two women accompanied their husbands on an expedition at the last minute (they were only supposed to sail as far as Chile). The experiment wasn't entirely successful. The two families fell out, and one of the women wrote later, 'I do not think that women belong in Antarctica'. After that, apart from a very brief visit from a Russian marine biologist, Marie Klenova, in 1956, there was nothing. By the 1950s a massive programme of science was starting on the continent, and it was strictly men only.

In 1969, Colin Bull, a glaciologist from New Zealand who had recently moved to the US, had been trying for nearly ten years to get a woman included in one of his field parties to the ice. The sticking point, time and again, was the US Navy. Though this was emphatically not a military operation, the Navy ran logistics for the US civilian programme, and they flatly refused to carry a woman on any of their transportation.

And then, Colin proposed an all-woman field party. He had been nagging for so long that the Navy was running out of excuses. They said OK, as long as he could find women with Antarctic experience. Right, well, one of them had worked on Antarctic rock samples. Check. Two of them had had to stay at home while their husbands went off to Antarctica. That was experience enough. Check. And the fourth could strip down motorbikes and carry heavy packs and, to be honest, by this stage, the momentum was there and the Navy men weren't bothering too much any more about the small print. The trip was authorised, and Colin received a short letter from one of the men who had spent the first winter at the Pole. It said: 'Dear Colin. Traitor!'

And so they went, a team of four (plus one penguin researcher who was working with her husband in a separate team). The right-hand man of the naval commander was in charge of making it work. 'We told these guys these women were scientists,' he said. 'They were married. Be respectful to them and don't smart off.' And, apparently, they didn't. The women reported that naval officers were almost exaggeratedly polite in their presence, and any rating who let slip a swear word got a tongue-lashing. Still, the news of the women's presence hadn't spread to everyone. Once in McMurdo, one of them noticed a man following her round, then later saw him sitting on a porch, crying. When she asked what was the matter he replied, 'I think you're a woman!' She reassured him that she thought so, too.[8]

The experiment didn't exactly open the floodgates, but it did create a chink for the many women who had been clamouring to get to the ice. Nowadays, the ratio of men to women at McMurdo isn't quite 50:50, but it's getting there. More significantly, the women are no longer just scientists. If you call for a carpenter or a locksmith or a mechanic, you are as likely to get a woman as a man.

There are still some women there who remember the old days. The year I visited McMurdo for the second time, in 2004–5, Sarah Krall was working in MacOps (McMurdo Operations), the control centre for much of the continent. She was the voice of Antarctica. Though she never left town, her voice was heard in every US field camp, on every helicopter and plane. It was Sarah who would raise the alarm if your helicopter pilot didn't call in after the agreed amount of ground time. She had to be ready for any emergency, and she also had to be able to juggle the constantly changing Antarctic timetables. (When you are handed a schedule for a plane or helo journey, it often comes with the rider 'the only thing certain about this is that it will change'.)

I went to catch Sarah at night, when most of the journeys

were over for the day, and the main radio traffic was from field camps reporting their weather, their numbers and their well-being: 'MacOps, MacOps, this is Beacon Valley. We have four souls on board and all is well.' All was conducted in the arcane language of high-frequency radio, where to distinguish numbers against the background buzz every 'nine' becomes a 'niner', where you say 'roger' and 'wilco' instead of 'yes' and 'I'll do that', and where every conversation finishes with the word 'clear'.

Then there was nothing but the background hiss from the HF, punctuated by occasional splurts caused by incoming cosmic rays (which Sarah calls 'cosmic raspberries'). She told me that the swishing sounds, like waves on a beach, were radio storms on Jupiter, and that low-pitched curtailed whistle was a passing meteorite.

Sarah first came south in 1985. She was the youngest of four kids, the baby, and being outdoors meant that nobody else would be bugging her. 'All I ever wanted was to have my own ideas and not be told what to do and how to do it.' In the end, her parents used to take Sarah and her dog out to the lakes near her home in Iowa and leave them there to kayak all day. By the time she applied to go to Antarctica, she had spent fourteen years as an instructor in the National Outdoor Leadership School. Still, it took two years for her to be accepted. (Her boyfriend was accepted in two weeks.)

And on her first view of the landscape, she was captivated. 'I felt like I had no place to put it,' she said. 'It was so big, so beautiful. I thought it might seem bare, but that b word didn't occur to me. Antarctica was just too full of itself.'

The US Navy was still running logistics then, and there were twenty-eight civilian women out of perhaps a thousand people. Of course that meant there was a permanent spotlight on every woman in the place. The men who didn't think women should be on the ice had no qualms about saying so. For many of them, if women could do this then it wasn't such a big deal, it didn't feel quite so heroic. But there were also, she said, plenty of men who were delighted with the changes, who loved having women as colleagues and friends.

'A few years in, a bunch of Russian men were coming through town,' she told me, 'and the National Science Foundation Rep, Dave Bresnahan, invited me to a cocktail party for them. "What is this?" I asked him, "An escort programme?" But Dave said, "No, Sarah. The Russians don't let women in their programme. I want you to talk to them, tell them what you do. Influence them."' She was clearly touched by this. 'It wasn't women fighting a world of men,' she said. 'It was women and men together, fighting the bigots.'

Over the years, Sarah had now worked in many different roles: in the field centre, as a mountaineer, as a helicopter technician, and she even operated the base's short-lived (but legendary) hovercraft. But her season as camp manager up on Mount Erebus was the one she most wanted to talk about, the one she still saw when she closed her eyes and thought about the ice.

Mount Erebus overlooks McMurdo and much of the rest of Ross Island. It is the most southerly active volcano in the world, the bottom part of the Pacific ring of fire, and one of the very few volcanoes that has a permanent molten lake of lava. It looks like a softly sloped mountain, draped with snow, except for the distant wisp of smoke that is visible whenever the cloud clears.

Erebus is high – nearly 13,000 feet – and Sarah's camp was within striking distance of its summit. One day near the end of the season she took a skidoo as far up as she could get and then hiked the rest of the way to the crater. When she reached the true summit and looked south for the first time, she burst into tears. 'I looked right down the peninsula and saw that tiny thing that was Ob Hill and to the right, the Erebus ice tongue looking like a chainsaw. And I thought, "How did I ever earn this? What will I have to do to pay for this?"'

It took her three hours to walk round the rim. The weather was cold but beautiful, and the air stank of rotten eggs. (She got 'plume cough', she said, from hiking in the cold with those sulphurous acid smells.) She stopped at the classic viewing point, to look over the rim. To the left, the east, was the lava lake, not a big red lake filling the whole crater but a patch of dark black crust with red lines running through it. Around it, the crater was littered with glossy black bombs that had been flung out by the lava, the size of a hand, a chair, a car.

'People talk about measuring the land, to get it down to size. I say no, don't do that. I use the land to take my measure. Am I competent enough to be there, to survive there? On a calm day at Erebus nothing seems more innocent. And then it throws up bombs with no warning. Kerpow! It's visceral. This land makes me feel small. Not diminished, but small. I like that.'

Her description was spellbinding. I couldn't understand why she was now prepared to bury herself in this dark building for the season. 'I love this,' she said, gesturing to the banks of switches and microphones in front of her. 'I don't know why. I guess I like feeling involved.' But then she shrugged and gave a rueful smile. 'I do miss being outside, though.'

Jules Uberuaga was another of the pioneers. A diminutive dark-haired firebrand, she stands all of five feet two inches and drives the heavy equipment, the big macho snow dozers used to build the runways and dig out trenches and level platforms. You might run into her in one of the bars; if you ask her nicely (and flirt with her a little) she might offer to take you for a ride in her beloved D7, a massive snow dozer that she has dubbed Trixie.

She will explain the best way to flatten a skiway, or dig out a building that is buried in snow without getting stuck yourself. She talks of the need to have a 'bubble in your ass' – an instinctive sense for when a surface is level. She will coach you in the subtle variations in angle and blade that will generate a neat roll of snow barrelling in front of you. And if you can maintain the roll without letting it break for the distance Jules sets, she will take a picture of you, standing in triumph on top of Trixie's roof.

Jules first came down here in 1979 when she was just twenty-four. There were few women in the programme and none driving heavy machines. One of her early jobs, out on the sea ice maintaining the airstrips, was immediately threatened when she was told that she couldn't use any of the men's bathrooms, which is to say she couldn't use any of the bathrooms. She was only saved when the servicemen let her use the facilities in their medical centre.

In her thirty-plus seasons she has heard everything you can imagine about why women shouldn't be in Antarctica in general and in a D7 in particular. But just as with Sarah, for all the men who protested there were always plenty of others ready to help her fight her corner. When she told one supervisor he was a 'fucking asshole' and asked if what he really wanted was to hit her, he screamed in fury: 'Did you call me an asshole?' 'Yes!' She shouted back. 'Well,' he replied, 'it's about time someone did!'

Now Jules is a veteran, as essential to McMurdo as the buildings and the furniture. 'I bet I've pushed more snow than any woman in the world,' she says, not in a boastful spirit, but as a simple matter of fact. Like the forams she has thrived in a world that looked well above her weight; and you sense when you speak to her that she has survived in part by building herself a protective outer shell.

The rest of the people here have also found their own adaptations to this strange way of living. They wear intricately crafted zipper pulls or carefully sculpted beards. They make spoof videos of classic sci-fi movies, or songs from *The Muppet Show*, exchange complex salutes, sit in the galley or the coffee shop arguing fiercely about Nietzsche or about game show hosts. Relationships form quickly and can break just as quickly. Some of the contract workers here have 'ice husbands' or 'ice wives', couplings that seem no less committed for the fact that they only exist when both partners are down here together. Even though the station has twenty-four-hour internet access, the outside world barely interferes.

The world of Antarctica, however, can make itself felt even for those who rarely get the chance to leave town. With regulation gear, the cold isn't so hard to deal with. But once in a while the winds will whip up into a 'condition 1' storm, in which visibility is zero and it's dangerous even to feel your way the few metres from one building to the next. During a condition 1 all outside travel is forbidden, and wherever you happen to be is where you have to stay. 'Hurry up and wait' people say to each other, with a shrug, wherever they are trapped. They will break out the playing cards, switch on the stove, and start one of the ubiquitous coffee machines.

This applies even more during the winter, when the permanent darkness falls, and the winds rise, and the cold gets into your bones. If winter storms sweep into town, you stay where you are. And if you're outside, you'd better be near a shelter.

The animals near here have learned this, especially the true Antarcticans, the ones that don't leave for the north no matter how bad things get. When winter approaches, Antarctica's Weddell seals stay in the water, trapped under an ice lid that can be metres thick, gnawing holes that are more like tunnels to enable them to breathe, spending months on end swimming, feeding and resting down there in the darkness, sheltered by the freezing water from the even harsher outside.

Nobody has ever seen these seals during the winter; the best clues for how they make their living come only in summer, when the females at least haul out to give birth, and moult and prepare again for the coming ordeal.

• • •

The sky was enormous, flecked with clouds that pointed like an arrow all the way to the horizon. Between them was bright blue sky and sunshine, a glorious day. I was driving one of my favourite Antarctic vehicles – a 'Mattrack' – which would be a perfectly normal red pick-up truck except for the triangular wheels. They made me laugh. From a distance it looked impossible, as if the wheels themselves should clunk awkwardly round. It was only when you were close that you noticed they were actually individual caterpillar treads, their triangular shape ensuring the best possible grip on the slippery sea ice.

The pace was faster than walking but not by much. I wound down my window, the better to see the view. The air that crept in was sharp but not unpleasant; the temperature couldn't be far below freezing and already I'd stripped off my parka.

A set of sea ice 'roads' branched away from McMurdo like a family tree. All were bright white, scored with skidoo or caterpillar tracks, and flanked on the right with a long row of flags, most red, a few green, fluttering from bamboo poles every few metres. The flags looked absurd, like overkill. Why did we need so many? But I knew that Ross Island could deliver a storm with a quick, casual sideswipe that would turn this big bright view into snow and confusion.

The weather forecast looked good but before being allowed to come I had to learn exactly what sized cracks I could safely cross, and how to pitch an emergency tent, in the lee of the vehicle, using ice screws that take painfully long to twist into the hard grey sea ice. I was also forbidden to travel alone, so beside me in the passenger seat was Mike from the heavy shop, whose name was next on the list of contenders to get out of town free, and who was quietly pleased at the outing.

We were going to meet Bob Garrott, a researcher from Montana State University who I ran into in the Crary Lab on his way out to the field. Bob worked at several different Weddell seal colonies, but he said we'd find him today at Turk's Head, the big blunt end of a rocky cape that juts out into the sea ice just beyond the Erebus Ice Tongue.

He'd already told me that right now, in early November, was the perfect time to study Weddells, because this was more or less the only time any of them came up out of the sea. Even so, we were unlikely to see males since they would only usually be on the surface when they had just lost a serious fight for underwater territory. But the females had to emerge to give birth, and we were entering prime pupping season.

Bob had also told me why he was so interested in Weddells. 'When animals are on the edge, stressed out, they come up with the most interesting survival strategies. I'm intrigued by the places and times where they have to go to the greatest lengths.'

Every so often, we saw a seal in the distance. They looked like slugs, fat and dark and lying utterly still. One advantage of the long American presence at McMurdo was that researchers had been studying and tagging the seals around here for nearly forty years. Any Weddell that you saw here was almost certain to bear a bright yellow or blue tag on its flipper. It could be jarring to see this overt sign of the presence of humans among animals that ought to be wild. But the tagging neither hurt the seals nor impeded them in their swimming, and it provided a spectacular database to trace how the most southerly mammal on Earth could make a living in a place that should surely be out of bounds to warm-blooded animals.

We parked the Mattrack and Bob came over to greet us. He led us off towards the colony, probing every so often with an ice axe. 'Whenever you're working around seals there will be cracks,' he said. 'Be careful and step where I step.'

As we walked, Bob explained that he was working now on population dynamics – how the seals lived and what sort of trade-offs they needed to make. All the other Antarctic seals live much farther north, in the pack ice, where stretches of open water form daily and air is easily had. Weddells are the only marine mammals to live in fast ice – where the sea covering is thick and there are very few breaks – and they have to go to considerable lengths to achieve this. It's hard being an air-breathing mammal in a place where air is scarce. They have developed special hinges on their jaws so they can open their mouths extraordinarily wide; with their inclined upper incisors they gnaw at the ice to keep breathing holes open throughout the winter. And they can hold their breath for up to an hour and a half before they have to find a hole, trumpet a warning that anyone else using it should get out of the way, and then surge upwards for that first fresh gasp of air.

Why should they go to so much trouble? Bob thinks it is so they can exploit scarce resources with much less competition. No other mammals are down here hunting the fish. And, perhaps more importantly, nobody else is on the hunt for seal pups. Up in the pack ice, killer whales and leopard seals prowl, but neither of these can live down here.

By avoiding these predators, seal pups have an unusually high survival rate compared to other similar creatures. But even so, only one in five of them will make it. Bob is interested in the details of this stark number. Who has a better chance than whom? What does it take to get ahead of the pack in the survival stakes?

Now we were walking among the colony proper, and Bob's colleague Mark Johnston came over to say hello. He saw me staring at two dead pups. They were pathetic scraps, one with its face planted in the snow, the other, even thinner, bent at an awkward right angle. 'That's part of the eighty per cent mortality right there,' he said. 'That little skinny one starved to death. It took about six days before he finally succumbed. It was sad. You wanted to do something like take him back to the hut and put him in a sleeping bag.'

I found this heartbreaking, but my head told me that even if they want to intervene, even if they could do something to save one of these creatures (which in the case of the pups was doubtful), they couldn't afford to. Bob and his team were studying what made the difference between life and death out here and any interference would hopelessly skew their results.

Their current study involved weighing both mothers and pups to see what effect their body size had on their chances of survival. They had brought in a weighing sled, towed behind a skidoo, and they were about to approach the first customer of the day.

Bob beckoned me over to where a seal was lying on its stomach. Just beyond was a pup so new to the world that its umbilical cord was still attached. It was a soft brown colour, small and slender against the inflated grey bulk of its mother. There were streaks of blood on the snow from the birth, and remnants of a placenta. Large seabirds were hovering nearby, waiting for the opportunity to scavenge this bounty.

As we approached, the pup moved nervously towards its mother, curling into her like a comma. A couple of researchers came up behind it and grabbed it with ropes. The pup looked tiny and fragile, and at first I couldn't understand why two of them were having such trouble dragging it over to the weighing sled. But when someone called out the reading I discovered to my astonishment that already it weighed half as much as I do. The mother followed, heaving her bulk clumsily over the ice, lurching on her flippers. She was calling like a wookie, her wails echoing off the cliffs. The pup's replies sounded distressingly like those of a crying baby.

And yet, the mother climbed cheerfully enough on to the sled to join her pup and someone called out the reading: she weighed more than a thousand pounds. Then mother and pup clambered off and lay placidly on the ice, just a few feet from the sled. It was astonishing. These gargantuan beasts had no land predators so they had no evolutionary reason to be stressed when they were up here on the ice. And that seemed to be enough to wipe out any memory of what seemed so anguished a moment ago.

But, then, I shouldn't be so quick to interpret the sounds Weddells make. They have twenty different vocalisations – everything from a chugging truck noise to high-pitched chittering. I have heard recordings of whistles, booms, gulps and chirrups. They emit alien whooshing electronica, sounds that shouldn't rightly come from any animal, let alone a furry one.[9] Bob told me that you could hear these sounds in the camp outhouse, resonating eerily up through the hole in the ice.

As he spoke, I was trying to take notes but my pen was giving me trouble. I scribbled irritatedly, wondering why it had so abruptly run out of ink. Mark glanced over. 'You're not using a pen are you?' he said incredulously. 'You can't do that out here – the ink freezes! Here.' He rummaged in his pocket and handed me a pencil. I noticed belatedly that the people noting down seal weights and dimensions were all using pencils. My issue gear was so warm, I'd forgotten that it was still sub-zero here.

Chastened, I left the team to their weighing and wandered off among the sparsely scattered colony. Near me, a seal abruptly poked her head through one of the two breathing holes, surrounded by slushy ice crystals. She blinked. Her eyes had a purplish bloom; they seemed to be all pupil, perhaps to help her hunt in the dark. Droplets of water dangled like beads on the ends of her whiskers. For a while she just hung there, inhaling deeply through her nostrils, holding her breath as if savouring the sensation, then releasing it with a snort. Each time she held her breath, I found myself holding mine, and then gasping on her behalf.

Suddenly she opened her mouth wide, called loudly, with a deep-throated 'coo-ee', disappeared from view, and then hauled out in a dramatic whoosh of water and slush. Even though she came out on the opposite side of the hole from me I leapt back in alarm and Bob looked over and laughed. 'Wahay! That was a thousand-pound rocket!' The call was to her pup, who responded by hurrying over and clamping to her invisible nipple like a limpet.

Bob told me that this urgent feeding process was the most important adaptation the Weddells had made to living down here, on the edge of survival. First the mothers store vast amounts of energy in their blubber – which is why they look so pumped up, as if their skin was ready to burst. Then they quickly dump an astonishing quantity of this into their pups. When they are nursing, even their blood is so full of fat it looks thick and creamy like a milk shake, and their milk itself is like warm wax. From just before birth to weaning, the mothers will lose nearly half their body weight in less than forty days. Bob pointed to a pregnant seal. 'Right now she looks like a fuel bladder. After she's pupped and weaned she'll look like a long thin cigar.' And the pups go the other way. They start at about seventy pounds, and within a month they will weigh five times as much.

The effectiveness of this process is crucial for survival. Pups are more likely to live through to weaning if they are born to a larger mother.[10] And the heavier the pup when it is weaned, the more likely it is to survive to adulthood.[11] I like this. Survival not of the fittest, so much as the fattest.

I also like the team's other key finding. For clearing those two hurdles of first survival through to weaning, and then to maturity, it helps immensely to be born to an older mother. The svelte young things of maybe six, which have only just started to breed, turn out to be less effective at raising pups than the ones that have been around the block a fair few times. The researchers think this is because the older the mother is, the fatter she is likely to be and the more energy she can pump into her offspring.

(The benefit disappears when the mothers get really old. In a reversal of middle-age spread, a twenty-two-year-old Weddell has such worn-down teeth from gnawing on years of sea ice that she is a less effective hunter, and hence tends to be less substantial than, say a fourteen-year-old.)[12]

The rest of the team was now struggling with a mother–pup pairing that didn't want to play. They had managed to drag the pup over to the weighing sled as bait, but it was thrashing and wailing. The mother had humped her way over to the sled but refused to get on. She rolled over, buried her nose in the snow, arched her back and then tried to get out of the way of the student blocking her path. Everyone was still and silent. Would she settle? No, she rolled away again. 'She's not going to do it. We'll leave them be.'

They released the pup to rejoin its mother and, once again, they were now both as docile as ever, the drama apparently forgotten. Still it seemed like hard labour, and I could see that every data point was dearly won. I asked Bob why he was prepared to put in so much effort.

'The best ecological insights come from living in the field and grunting around from day to day,' he said. 'You can use satellite images all you like but you don't really learn things until you are watching them every day saying "why are they doing that?"

'Most of the time on the ice they are just sleeping and nursing. But if you look at the same animals again and again you start to realise – that's a good mom, that's a bad one, that's one's mellow, that one's psycho.

'Sometimes if you look down a hole with a parka over your head to block the light you can see them below the ice, perfectly at home in water that's at 28°F. I'm constantly amazed at how they survive in such a crazy environment that shouldn't support life. And you know what? They don't just survive there. They flourish.'

On the way back to town, I thought about those mothers urgently fattening up their pups against the coming winter, and how in spite of this four out of five of the pups would die. That McMurdo mantra may be said ironically but it was also true: this *is* a harsh continent. And I was beginning to feel that we humans, the newcomers, could learn a lot from the creatures that had spent thousands or even millions of years figuring out what adaptations it takes to flourish here.

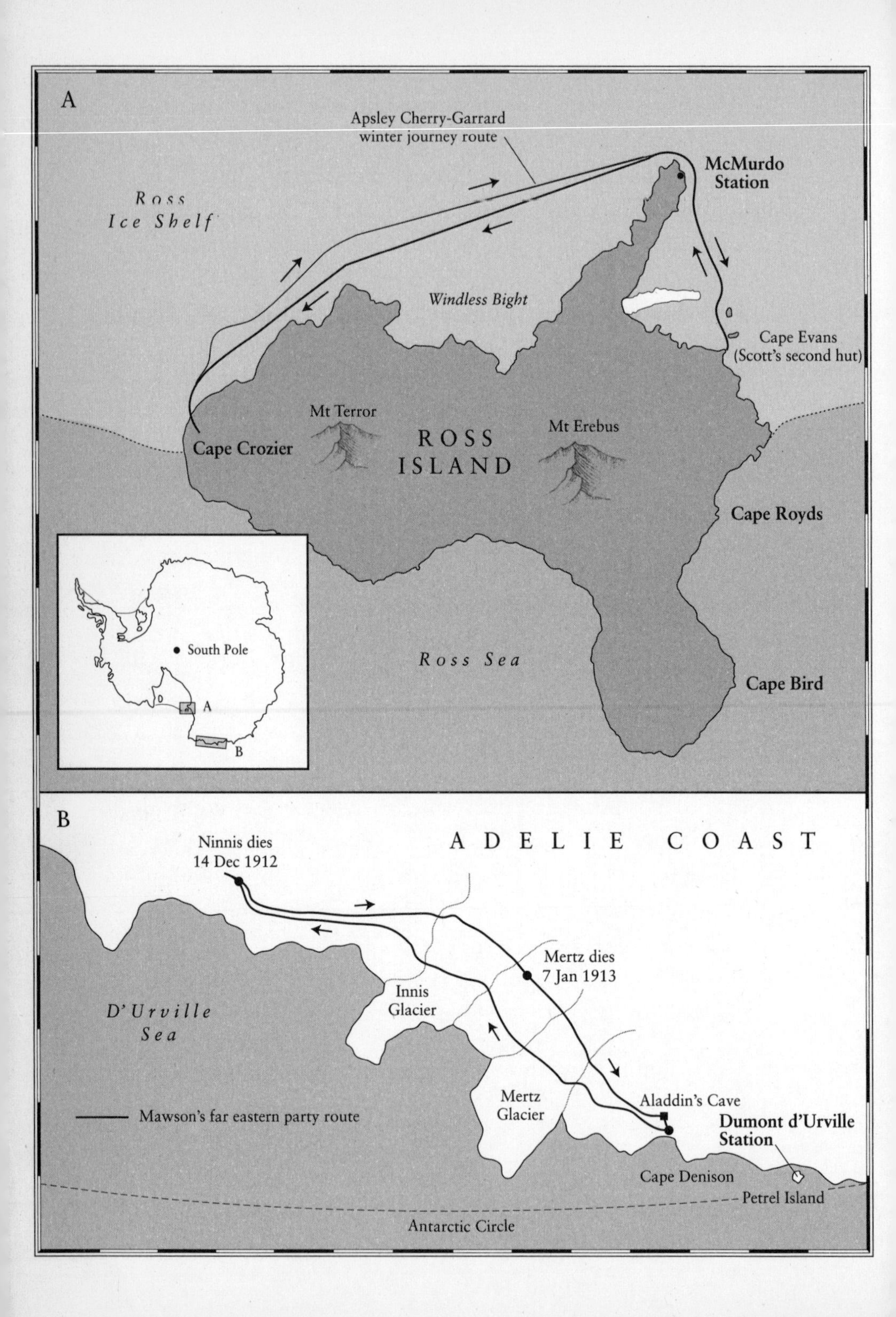
A
Apsley Cherry-Garrard
winter journey route
McMurdo
Station
Ross
Ice Shelf
Windless Bight
Cape Evans
(Scott's second hut)
Mt Terror
Mt Erebus
Cape Crozier
ROSS
ISLAND
Cape Royds
South Pole
A
B
Ross Sea
Cape Bird
B
Ninnis dies
14 Dec 1912
ADELIE COAST
Mertz dies
7 Jan 1913
Innis
Glacier
D'Urville
Sea
Mertz
Glacier
Aladdin's Cave
Dumont d'Urville
Station
Mawson's far eastern party route
Cape Denison
Petrel Island
Antarctic Circle

2

THE MARCH OF THE PENGUINS

David Ainley looked like an ageing surfer dude, or a mountaineer who had spent a little too much time squinting into the sun and wind. He was in his early sixties and had been coming to Antarctica for ever. He had an untamed shock of white hair, a heavily tanned face, a moustache that he keeps when he's off the ice, and a beard that was here just for the season. I had been warned that he wasn't so good with people. 'Taciturn' was how some had described him to me, and 'a bit wild'. He spends as much time as possible out here in his field camp, among his penguins, and as little as he can back in Mactown.[1]

David was Californian, a biologist from an ecological consultancy in San Jose. He spoke slowly and hesitantly, as if he couldn't quite remember how you're supposed to talk to other humans. Sometimes he put invisible inverted commas around his words and pronounced long ones in an exaggerated way as if he were making some kind of joke. He often was making some kind of joke. I liked him immediately.

He pulled on his coat as I entered the main research tent.

'Come on then,' he said. 'Let's see how many smiling faces greet us.'

'Huh?'

'Penguins are always smiling. They have no self-doubt.'

We stomped out of the tent in our bunny boots and headed down towards the sea. David's camp at Cape Royds was a short helicopter ride from McMurdo, on the westernmost tip of Ross Island, and was home to a colony of Adélie penguins. They are classic cartoon creatures, knee-high, with black heads, flippers and backs, white throats and chests, and a bright white ring around their eyes. They are pathologically busy, packing their entire breeding cycle into the brief Antarctic summer. Adélies are also cute and comical. And everybody, but everybody, loves them.

I, however, did not. Even before I met them I was already tired of penguins. From the moment I started to talk about Antarctica to my friends and family I began to receive a mountain of penguin paraphernalia. There were penguin T-shirts, penguin cards, penguin jigsaw puzzles, cups, mugs, glasses, playing cards, a penguin apron, penguin pyjamas. For birthdays, Christmases, or for no particular reason I received penguin backpacks, pencils, rulers, scarves, gloves, big furry penguins, small furry penguins, penguin place mats and cutlery. When I outlawed penguin presents they still came in. 'It's just a small one. I couldn't resist.'

Well, I could, quite happily, and I was determined to resist the charms of the real Antarctic thing. Penguins are the clichés of Antarctica, annoyingly cute icons of a continent that is otherwise wild and vast and mysterious. I saw them as our way of diminishing the ice; we anthropomorphise them, personify them, imagine them to be amusing little people, and in the process we bring the continent down to a manageable human scale. I hated that idea. So although I love animals in general, I had told my friends and I had told myself that I would not, repeat not, fall in love with these creatures. I would write about penguins because there was interesting science to tell. That was all.

It was a gorgeous day, barely below freezing, with a bright sun reflecting off the sea ice that was crammed up against the shore. Most of the ground was bare volcanic rock, and Mount Erebus's bulk dominated the scene, topped with its customary cloud.

David told me that Adélies choose to settle their colonies in rocky places – to keep their eggs safe from ice – and those that are also close enough to the sea to enable them to fish for food. But although we were looking down on to McMurdo Sound, there was no open water between here and the horizon, just endless sea ice that had been squeezed into ridges and scattered chunks as if a giant child had thrown toy blocks out of its pram. A fat Weddell seal had hauled itself out of a crack in the distance. Seven Adélies were making their arduous way back, stumbling and falling flat in the gaps between ice chunks, scrabbling with their flippers to pull them through. It looked hard. There was still a fair way to go before they reached the rocks, and then another stiff climb uphill to the colony proper.

We passed a small pond half clear of ice with a running stream spilling out over the rocks, the edges scattered with white eggshells. And then we reached the colony, and the noise raised itself from a rumour to a roar; it sounded unhinged, like a cackling orchestra of kazoos. The fishy guano smell was noticeable, but not nearly as strong as I had been expecting. Although the nests were densely packed on the ground, I supposed in this big open spot there was enough wind to whisk the ranker smells away.

By contrast with the bustling birds heading to and from the sea ice, the ones on the nests were placid, and even listless. Every so often one stood up, stretched and flapped its wings, revealing the bare pink patch on its belly that fitted neatly over the eggs, skin to shell, to ensure that maximum warmth reached the chicks inside. David told me that they would usually lay two eggs apiece but this year there had been quite a few single clutches.

A skua landed and started prowling. It looked like a seagull but larger and brown with a wickedly curving beak like a hawk. I'd seen skuas at McMurdo, and been warned that they would snatch a sandwich from your hand if you let them. They are scavengers, always out for an opportunity to harass, bully or steal. This one was clearly eyeing up the eggs. The nearest penguins stretched their necks menacingly like guard geese ready to hiss. They shuffled round, keeping their eggs out of sight beneath them, maintaining eye contact with the enemy.

'Why doesn't the skua just attack?' I asked.

'It's afraid of the penguins, with good cause,' said David. 'Skuas may stand as tall as penguins but they are all air and feathers. They only weigh maybe nine hundred or a thousand grams. The penguins are much denser – they can weigh seven or eight kilograms. And those flippers are very hard. A whack from one of those and a skua definitely remembers it.'

That doesn't stop them from looking for a quick thieving chance. While our skua was getting nowhere, a burst of indignant squawking erupted to our right and another bird flew overhead, barely able to hold the outsized egg in its beak. The penguins settled back down on their nests. There was no sign of where the egg had been taken from. Everyone seemed resigned.

Now in mid-December, the Adélies were almost halfway through their race against time. Each year they must pack every relationship stage of meeting, wooing, mating, hatching, rearing and weaning their young into the few short months of the summer season.

They come here in early November at the start of the southern summer and hastily reconvene with last year's partners. There is little time wasted on courting niceties or on excessive fidelity. If you're a day or two late, your former mate will already be on to someone else. The first eggs come about a week later, and by two weeks most of them are laid.

The penguins are fat when they first arrive, having stocked up on fish for the season. As soon as she has laid her eggs the female heads back to open water to replenish, and the male incubates the eggs and keeps the skuas at bay. If he's lucky, she'll be back within two weeks to relieve him.

For the first few weeks after hatching, at least one parent will stay with the chick. But when the chicks are bigger and more demanding, both parents will go off to feed at the same time, saving some fish for themselves and regurgitating the rest into their offsprings' gaping gullets. By now, the chicks will probably be in a crèche, kept safe by adults that are hanging around or still tending nests. When they are seven or eight weeks old the chicks will lose their soft brown down to reveal a grown-up blue and white penguin suit that will soon turn black. From the first week of February they will head for the sea and be on their own.

Released from their duties, the parents will feed voraciously for a few weeks to regain their fat reserves. Then they will haul out on sea ice floes and moult. This is apparently the one time these little penguins do not smile. David told me that when they're moulting they don't like to be touched. They just sit there, he said, scowling, wanting no other penguins anywhere near, until they have lost their old plumage and grown a new one. And then, as the first fingers of winter touch the continent, they will head north. Not too far, though, for they are true Antarcticans. They may retreat a little, to the edge of the open water to wait out the winter, but they never leave the ice. And then, at the start of spring, they return loyally to the exact same nesting spot, to start all over again.

At least, that was what normally happened. But the past few years had been challenging ones for the Adélies of Ross Island. In

March 2000 a massive chunk of the Ross Ice Shelf broke off to create one of the largest icebergs ever seen. Though it subsequently broke into several pieces, the biggest of these – called B15a[2] – still measured more than a hundred miles long, and was larger than the state of Delaware.

B15a wedged itself across much of the mouth of McMurdo Sound, blocking the route back from the penguins' winter homes to their summer nesting spots with a giant white cliff. The only options were to turn left and head to the massive colony at Cape Crozier, on Ross Island's eastern side, or to turn right, double back on yourself for miles, and then round a corner and find the way here to Cape Royds. Crozier was by far the easier route.

So what did they do? Thanks to the activities of David and his colleagues, many of the penguins in both colonies wore bands on their flippers, marking who they were and where they were born. That turned out to be the perfect opportunity to find out how loyal they really were to their place of birth. And much to David's surprise the answer was . . . not very. Royds banded birds had been showing up at Cape Bird, thirteen miles away, and even at Cape Crozier, which was forty miles from where they were supposed to be. Very few had gone the opposite way.

'This has rewritten the book on immigration and emigration,' said David. 'Adélies were supposed to be highly philopatric – religiously returning to the colony where they were born. But now we know that their behaviour is much more pliable than we thought.'

That was great – in a way. But the megaberg brought a darker side, too. The sea ice would normally break up during the summer, but the gigantic cliffs of B15a had encouraged it to stay around. Adélie colonies are usually within a kilometre of open water, to enable the parents to forage for food quickly and easily, since they can swim much faster than they can walk. But thanks to the iceberg, sea ice now stretched farther than we could see.

And the Adélies were suffering. 'There are so few birds,' said David. 'You'd normally have non-breeding birds or practising juveniles to chase the skuas away, but there's nobody this year younger than four, and lots of birds who had been trying to breed have now abandoned their nests.'

He went off to check who was still here, stepping carefully between the nests, stopping to make pencil notes in an orange notebook. Then he beckoned me over in delight to see the first chick of the season. Above the background cackling there was suddenly a racket that to my inexpert ear sounded like some kind of territorial challenge. But David beamed and said, 'That's a returning partner ready to take over.' Much cawing and squawking followed; the two penguins threw their heads up into the air and opened their beaks operatically wide, and then snaked their sinuous necks around one another, to the left, and the right, and the left, and the right. Then, in a heavily choreographed move, the one on the nest stepped aside and the new one immediately shuffled in to take over. The eggs beneath were visible for a bare second or two; they were a little larger than duck eggs and just off-white. The returning female settled down, stood up, shuffled the two eggs around, and sat down again. Her relieved mate hung around for a few moments, picking up a few stones and adding them to the nest.

'Stones are penguin currency,' David said. 'They are a prime factor in an Adélie penguin's self-esteem.' They are important because they keep the eggs high and dry, free from any meltwater that might run by. If the temperature just topped the freezing point, snow could melt. But the water would quickly freeze again, and if it were touching an egg, the chick inside would freeze, too. The bigger your pile of stones, the safer your eggs, and the more you can thrust out your chest and trumpet and crow to the penguins around you. Stones are the penguin equivalent of designer brands or fast cars. They are the outward, in-your-face signs of success. 'Once penguins have accumulated a big pile they become very possessive. They will steal stones and squabble over them. There's no chance to replace stolen stones when the skuas are around – you have to sit tight and wait for your mate. So every time there's a nest relief the outgoing bird is supposed to find new stones to add to the pile.'

Our newly relieved male made a few half-hearted efforts. Normally he would stick around for an hour doing this, but this one was woefully skinny beside his plump partner. After just a few minutes, he turned and headed off to the slope that led down to the sea ice.

We watched him go. 'This experiment with the iceberg was made to order,' said David. 'But now it's getting old.'

Between David's tent and the penguin colony, a wooden hut rose incongruously on the rocks. Its outer walls had been bleached blond by the sun and the scouring effect of dirt and wind. Inside, as for the other Antarctic huts of the heroic age, it looked almost new, but the furnishings were old-fashioned enough to give it a historic air. There was a dual portrait on the wall of King George V looking regally to the side, and his consort Queen Mary staring out into the hut. On one side was a wood-burning stove and oven, with a metal chimney. The shelves still bore slightly rusty tins of food whose antique lettering declared their contents: 'Kippered Herring', 'Pure Preserved Cabbage', 'Irish Brawn'.

This last could also have aptly referred to the leader of the expedition that built this hut, the boss himself, Ernest Shackleton. Shackleton was born and initially brought up in Ireland, and, although he moved to England at the age of ten, to a private school that raised true sons of the British Empire, he never quite lost his Irish brogue. His snooty fellow pupils mocked him for his background, but he knew how to hold his own. One friend said of him: 'If there was a scrap, he was usually in it.'[3]

Back in 1902, Shackleton had been part of Captain Scott's *Discovery* expedition, for which the first hut had been built at Hut Point. Scott had chosen Shackleton as one of the two men to join him on the first ever attempt to reach the South Pole. But the journey was poorly planned and the outcome woeful. All three explorers ended up suffering from scurvy, and they barely managed to drag their overburdened sledges nearly five hundred kilometres from base before having to turn back. Scott knew that Shackleton was liked and respected and somehow seemed a threat to his authority. He was also looking for someone to blame for the fiasco. He declared that Shackleton was medically unfit for duty and ordered him home.

This new attempt on the Pole, which Shackleton had cobbled together using a tiny ship called the *Nimrod*, was his response to that earlier humiliation. It was vital that he succeed, particularly as Scott had been livid when he heard that Shackleton was now striking out on his own.

The contrast between the two men was marked. Scott was a commissioned officer in His Majesty's Royal Navy; Shackleton – whose physician father could not afford to send him into the Navy – had learned his seamanship in the less prestigious Merchant Navy. Scott was formal; Shackleton was charismatic. Scott drew a strict division between officers and men; Shackleton had an open-plan attitude to the architecture of leadership.

So much was clear inside the hut that Shackleton built here at Cape Royds, where he arrived in February 1908. Every man was treated equally, and given his own space. The edges of the hut were divided into two-man cubicles, 400 square feet, each with a suitable nickname. One was so immaculate, with such high-brow books along its shelves, that it was dubbed '1 Park Lane', then the smartest address in London; another was crudely called 'The Taproom' since one of its occupants suffered chronically from diarrhoea; still another, which belonged to the two scientists and contained a jumble of bizarre instruments and devices, was 'The Old Curiosity Shop'.

Shackleton was the only occupant of the hut who had a room to himself. But that was as much because he knew that the men sometimes needed to kick loose away from their leader as because he wanted privacy. Throughout the winter of 1908, the hut was a happy one. The boss may have had a quick temper, but it passed just as quickly. More importantly he had the knack of making everyone feel they were uniquely essential to the mission.

With the return of summer, on 29 October 1908, in brilliant sunshine and under a cloudless sky, Shackleton set out in pursuit of his dream. He was accompanied by three chosen companions, a support party, a motor tractor and a set of Siberian ponies each gamely pulling a load.

The motor tractor was soon struggling on the uneven surface of the great Barrier, and the ponies did not fare much better. But still the explorers marched on. The support crew deposited food and supplies for the return journey, and then on 7 November they left to return to Cape Royds. Shackleton and the three remaining men passed Scott's previous furthest point south with ease, and soon they were witnessing something no human had ever seen. They had reached the end of the flat white plain of the Great Ice Barrier, and in front of them rose a range of magnificent mountains. The four men climbed up one of the smaller mountains, which they named Mount Hope, to spy out the land. 'There burst upon our vision an open road to the South,' Shackleton wrote. They had discovered a vast gleaming glacier, which he called the Beardmore Glacier after one of the expedition's wealthy sponsors. This would be their staircase south, up on to the great plateau of the East Antarctic Ice Sheet.

However, as they laboured up the glacier, and began to suffer from the high altitude they were attaining, Shackleton realised that their progress was worryingly slow. He had calculated their food based on achieving nineteen miles a day, but they were barely managing five.

Up on the plateau, the conditions grew worse. On Christmas Day one of the party, Frank Wild, wrote in his diary: 'May none but my worst enemies ever spend their Xmas in such a dreary God forsaken spot as this. Here we are 9500 feet above sea level, farther away from civilization than any human being has ever been ... with half a gale blowing, and drift snow flying, and a temperature of 52 degrees of frost.'[4]

After a hearty Christmas dinner, Shackleton surveyed the remaining food and decided that they would have to cut their rations, making each week's food last ten days. 'It is the only thing to do, for we must get to the pole come what may.'[5]

The explorers of the heroic age mainly survived on tea, cocoa and pemmican – an unappetising mix of dried meat and fat that could be reconstituted in a stew called 'hoosh', often with dried biscuits crumbled in. Every scrap mattered. They even devised a ritual called 'shut eye' for allocating the portions: one person would turn his back and another would point to each of the servings of hoosh saying 'whose?'.

And as the journey grew longer and the hoosh became thinner, the explorers were also increasingly prone to food dreams that were more like nightmares. They would find themselves, perhaps, at a food stand or a feast, with fresh bread and buns and chocolate and roasting meat. If they were lucky they at least got to taste it in their dreams; if unlucky, they woke night after night just as they were raising it to their lips.

The slope continued to rise, and the men continued to

struggle. The air was not just high, but also dry; with limited fuel they could not melt much snow and they were becoming dehydrated. Shackleton recorded how, after every hour of pulling, they threw themselves on their backs for three minutes to recover.

On 2 January 1909, they passed the previous record for the highest latitude achieved at either Pole. Still they ploughed on, as their food supplies diminished. Finally, on 9 January they passed through the barrier of 89° and found themselves within a hundred miles of the Pole. It was so close now that Shackleton could almost smell it. He knew that the party could reach their goal. But he also knew that there was not enough food left in their sacks to sustain them for the journey home. If they continued now, they would certainly perish. Survival mattered more than glory. And so he did an extraordinary thing.

He gave the order to turn back.

Even now, they had cut their journey very fine. Their food stores were so low that they barely reached each depot before the previous cupboard was bare. The men were exhausted, starving, scarecrows. They were racing against their own broken bodies. 'We are so thin that our bones ache as we lie on the hard snow in our sleeping bags,' Shackleton wrote.[6]

They were also racing with time. In their weakness they were moving so much more slowly than Shackleton had expected that they risked missing the ship that was due to take them home. When one man collapsed with thirty-three miles to go, and only thirty-six hours before the ship was due to sail, Shackleton left him in the care of the remaining team member and struck out with Wild, carrying nothing but a compass, sleeping bags and some food. The two men marched throughout the night and the next day, finally reaching Hut Point at 8 p.m. on 28 February. But the hut was dark and a note nailed to the door informed the readers that the ship would wait until 26 February, and after that would sail.

The men spent a disconsolate night, but they still hadn't given up hope. The next morning they set one of the outbuildings alight, and raised a signal flag over the building. And, praise be, in spite of the message, the *Nimrod* had not yet sailed. It saw the signal and steamed to the rescue.

When the ship arrived, Wild collapsed on board, but Shackleton hadn't finished yet. Two of his men were still out there on the Barrier. He pulled together a rescue party and announced he would lead it himself. I can't imagine how tired he must have been, how ready to fall, how relieved to be safe in the relative civilisation of the ship. But he wasn't even tempted to stay and let others take over this one last act. He didn't do it with any fanfare or in a showy fashion, but because he was a leader, and that's what leaders do.

Shackleton's wife described him once as 'a soul whipped on by the wanderfire'.[7] Perhaps that is true, but he was also whipped on by the drive to lead. Back in England, between expeditions, he seemed somehow diminished, a wide boy permanently full of foolish schemes to get rich quick. But on the ice he was magnificent.

There is a tale, perhaps apocryphal,[8] that Shackleton had placed an advertisement in an English newspaper seeking crew for his *Nimrod* voyage:

> *Men wanted for hazardous journey. Low wages, bitter cold, long hours of complete darkness. Safe return doubtful. Honour and recognition in event of success.*

Throughout his subsequent Antarctic endeavours, Shackleton never once achieved what he had formally set out to do. But he nonetheless gained his share of honour and recognition. He went on to perform daring rescues, had extraordinary adventures, and – in spite of the advertisement – never lost a single man under his direct command.

Within three years of Shackleton's abortive attempt, two different expeditions had wiped out his southernmost record and succeeded in reaching the Pole, although one of them would not survive the journey back.

'Better a live donkey than a dead lion,' Shackleton had observed to his wife. But he was no donkey. Turning for home, less than a hundred miles short of certain glory, was an act of extraordinary courage – one of the bravest things any Antarctic explorer has ever done.

Roald Amundsen, the eventual conqueror of the Pole, understood this better than most. In his later account of his own expedition, Amundsen said this: 'Ernest Shackleton's name will always be written in the annals of Antarctic exploration in letters of fire.'

Beyond Shackleton's hut, away from the main colony, lay a hollow between two small hills, where perhaps forty penguins were lying placidly on their nests inside a corral made from a green mesh fence that was thigh-high. This was a sub-colony, which David was studying separately from the rest. At first I thought the penguins were trapped, but then I saw a weighbridge, an archway over a metallic grey mat, piled around with rocks.

This experiment was all about food. 'Adélies are just bundles of energy,' David said. 'They keep going forward. Unless they're sitting on eggs they just don't stand still. It takes a lot of food, though. They definitely eat a lot of food.'

He told me the birds there each had a chip inserted under their skin, the same sort that people use to identify their cats and dogs. The whole thing was ingenious. When a penguin entered the bridge, it cut through an optical beam that switched the machine on. A magnetic field in the hoop overhead activated the transmitter in the chip that broadcasted the penguin's ID. Beneath the mat, an electronic scale measured its weight, and the bird then cut through a second beam so you knew whether it was entering or leaving.

'We weigh them in and out,' David said. 'When they're feeding the chicks, we calculate how much food they give by comparing the adult's arrival weight with its leaving weight. Doing that by weighing chicks causes too much disturbance.'

The penguins eat a combination of fish and a shrimp-like creature called krill. The researchers know this, David told me, because in the past they spent a lot of time making the penguins regurgitate.

'You pump them full of water, then you tip them upside down,' he said. 'It would take three goes to empty their stomach and we only did it once, so we left them plenty of food. We don't do it any more. It was distressing for both the researchers and the animals. It violated their sense of self.'

'Their sense of *self*?'

'They definitely have a sense of self, an aura of penguinhood. We try not to do anything that another penguin won't do. They behave towards you as if you're a large penguin. If you come close to their nest they will treat you like any penguin and peck you. Their beaks are very sharp and they whack you on the shins with their flippers. But if you pick them up, no part of their normal life is in the air. When we attach the transmitters we try to trap them between our feet. We'd only pick them up as a last resort.'

'What are they like to touch?' I asked him.

'If they're fluffed out, the feathers are soft, but if they're in sleek mode, they're almost like scales,' he said. 'They're very vigorous and extremely strong and they squirm a lot. They're mostly bone and muscle. The bones are solid, unlike other birds, and the muscles are huge for all that swimming and walking so they're really hard to hold. You have to grab both of their feet, tuck their head in your armpit so their eyes are covered, and hold them like a football.'

I risked a more personal question. 'Why do you do all this?'

'I dunno, I'm just interested in seabirds and the ocean and the ... um ... what you might call the romance.' He was warming up now. 'It's the idea of this vast ocean with these warm-blooded creatures that are pretty much in the same boat as some humans except they've figured things out a little bit better. In a way humans have totally trashed the oceans, whereas birds have solved it because, you know, they fit in. They don't try to change it.'

'How have humans changed it?'

'Well, we've removed the highest predators from all the other oceans. The whales, seals, cod, pollack, tuna, swordfish, sharks are all gone. If you have predators, there are long-lived creatures that eat the surpluses and coast during the troughs of food supply, so you get a stable system – one that doesn't swing so wildly from one extreme to the other.'

So according to David, because the Ross Sea still had birds and whales and seals and predatory fish, this was the only place on Earth where the ocean was behaving as it was supposed to.

'It's truly wild,' I suggested.

'Yes, it is wild. It's all out there in full view. There are no secrets. Penguins can't hide and they don't question anything. But you can ask them questions, and if you're creative enough you can find the answers.'

'Do you like the other kind of wildness – the elemental kind?'

Again, he paused to consider the question.

'I don't want dangerous. I'm not looking for an adrenalin buzz,' he said. 'But I do like Cape Crozier, which is famous for its wind. The weather there is really localised. You can sit in a place where it's calm and see a raging hurricane, just a couple of miles away; it's like a grey cloud zipping over the Ross Ice Shelf and turning the ocean into a white froth.'

'Have you ever been out in a bad storm?'

'Yes. A couple of times. You can be standing there in total calm and suddenly there are seventy-knot winds knocking you off your feet. One time, when I was a grad student, a storm had been blowing for like three days. When the winds finally stopped I donned my gear and ran down to the beach to check how the penguins were doing. If the storm is too strong they start to fly, and can get broken bones.

'As I was down there, the wind came back, and it was a total white-out, blowing a hundred knots. I couldn't really walk but I could crawl and I knew the sub-colonies of those penguins pretty well. We had a little observation hide, about the size of a telephone booth, so I found my way to that on my hands and knees. It had a sleeping bag and C rations – the food that army guys eat. I wouldn't touch the canned stuff – it's repulsive – but I ate all the cake. I was in there for about thirty-six hours. Then the whiteout stopped, though the wind was still blowing a hundred knots and I crawled back to the main hut about a kilometre away and . . . lived happily ever after.'

'Were you scared?'

'No, just bored.'

'What about the three other people in the hut? Were they worried about you?'

'Yep.'

'But they didn't come to find you?'

'They couldn't see.'

I digested this for a moment. You are trapped by fierce winds in a box the size of a telephone booth for a day and a half, eating nothing but cake, while your friends can't come and find out if you are alive or dead because *they can't see.* The air is so thick with snow and wind that there is nothing anyone can do but wait. This is truly not a continent for the impatient.

Three members of Scott's expedition to the Pole had their own horrifying experience of the winds of Cape Crozier, but in their case they went in the winter. I asked David if he knew the story.

'Oh yes,' he said. 'I read it while I was at Crozier. Those guys were out of their minds.'

He stopped, as if considering fairness, then added: 'They had no idea what they were getting themselves into. McMurdo is completely different because of the way the wind works around here. When we radioed in, people would never believe us that it was blowing a hundred and forty knots at Crozier. In McMurdo it was a nice day. Those guys had no idea what they would find. If Birdie Bowers hadn't been there to build that rock igloo, they would all have died.'

20 July 1911, Cape Crozier

> *I do not know what time it was when I woke up. It was calm, with that absolute silence which can be so soothing or so terrible as circumstances dictate. Then there came a sob of wind, and all was still again. Ten minutes and it was blowing as though the world was having a fit of hysterics. The earth was torn in pieces: the indescribable fury and roar of it all cannot be imagined.*[9]

This must have seemed like the end. Apsley Cherry-Garrard and his companions, Bill Wilson and Birdie Bowers, had suffered almost unimaginably in the three weeks it had taken them to travel to Cape Crozier. They had nearly died so many times, nobody was bothering to keep score. And just as they had finally reached their goal, built themselves a secure stone igloo, and settled down for an attempt at rest, they had been hit by one of Crozier's now legendary storms.

There was worse to come. The next thing Cherry-Garrard heard was a cry from Bowers. 'The tent has gone!' The three men had pitched their tent in the lee of the igloo, but the furious winds had ripped it from its moorings. They were more than 100 km from the hut that was currently sheltering Scott and the rest of the expedition, back at Cape Evans. Without their tent, in pitch darkness and the coldest temperatures any human had yet experienced, they now had no hope of returning.

The three men had come south as part of Scott's team for his forthcoming attempt on the South Pole. During the winter of preparations, Wilson, a fervent naturalist, asked Scott's permission to journey out to Cape Crozier and study the emperor penguins there. The reasoning seemed sound. Wilson believed – wrongly as it turned out – that emperors were among the most primitive of birds, and that their embryos might enable him to trace evolutionary links between birds and reptiles.

At the time the only known rookery of emperors was at Cape Crozier. And when an earlier expedition had sailed there in the spring, the chicks had already hatched. Clearly, emperors must incubate their eggs during the winter. If you wanted the embryos, you'd have to go then.

Scott gave his permission, Cherry-Garrard and Bowers volunteered to accompany Wilson and the three men cheerfully made their preparations. Nobody had ever travelled in an Antarctica winter. This would be exciting. But when they set off on what Cherry-Garrard called 'the weirdest bird's nesting expedition that has ever been or ever will be',[10] they had no idea what was to come.

First they were woefully unprepared for the cold. The temperature plummeted beyond their imaginings, bringing them blisters of frostbite if they dared to remove their gloves or expose any flesh for an instant. Their tents were just warm enough to begin the thaw, rendering their sleeping bags clammy and their clothes damp enough to freeze even more solidly when they re-emerged. One morning, as Cherry-Garrard left his tent, he stood up and lifted his head to look about. It wasn't for long, just ten or maybe fifteen seconds. But that was enough for his clothes to become completely rigid, trapping him painfully upright for the next four hours of sledge hauling. After that, all three men were careful to bend over into a pulling position the moment they went outside.

On 6 July, peering at their thermometer by the light of a flickering candle, they read a temperature of -77.5°F, the lowest ever recorded. This, said Cherry-Garrard, was the day he discovered that records are not worth making.

The nights should have brought some respite. But the men shivered so painfully and uncontrollably in their sleeping bags that they feared their bones might break. Cherry-Garrard wrote that the two worst jobs of the entire enterprise were first getting into the bag, and then having to stay there for six hours. 'They talk of chattering teeth,' he said, 'but when your body chatters you may call yourself cold.'[11] The call to wake and start the day's pulling came as a blessed relief, but, in the darkness and fumbling cold, it still took five full hours just to strike camp.

'I don't believe that minus seventy temperatures would be bad in daylight,' Cherry-Garrard wrote, 'not comparatively bad, when you could see where you were going, where you were stepping, where the sledge straps were, the cooker, the primus, the food; could see your footsteps lately trodden deep into the soft snow that you might find your way back to the rest of your load; could see the lashings of the food bags; could read a compass without striking three or four different boxes to find one dry match.'[12]

Conditions were worse than any of the men had ever dreamed of, but nobody wanted to be the first to say so. The expedition leader, Bill Wilson, repeatedly asked the other two if they wanted to return and each time they said no. Later he stopped asking and merely apologised, again and again, for the horrors he had led them into.

Still they trudged on, across the white emptiness of Windless Bight where the snow was so cold that it was like pulling over sand. The men could no longer drag their sledges in tandem. Instead, they had to relay them. Fumble and fasten your harness; heave through the cold and darkness for one mile; unfasten yourself; trudge back; hook up to the second sledge; heave, trudge, unhook, repeat. And all by the light of a naked candle, in temperatures that would freeze your soul.

Next came Terror Point, where the sea ice crams into the island creating mountainous pressure ridges over which they hauled their sledges, one at a time, up, over, down the other side, this one first and then back for the other. And then there were the crevasses. It was impossible in the darkness to see the snow bridges that draped them. All you could do was crash through, hope your harness would hold, climb out, crash through again, and hope and climb and crash and hope again. When they finally reached the emperor rookery, their bodies and minds were all but destroyed.

But at Birdie Bowers's insistence they built themselves a stone igloo. And then, aware that a storm was coming, they hastened down to the rookery and collected five eggs, cushioning them in their mittens. Poor Cherry-Garrard took two, but smashed them both. It wasn't just the darkness; he was also hopelessly short-sighted and the cold meant that he had no chance of wearing his glasses.

It was when they returned to their igloo, three remaining eggs in hand, that the storm struck. It shattered the canvas roof of their igloo. Rocks and snow rained down on them and the wind tore through like an express train. Beaten down by the force of the hurricane, they cowered in their sleeping bags, sucking on snow for water. But somehow, for the two long days that followed, they clung on to some shreds of their selves. They huddled together; they said 'please' and 'thank you'; they sang hymns, feebly, against the roar of the wind.

And then, when the storm finally died down, they staggered out of the igloo and went to look for their tent. I still can't believe they did that. In the half-light, in the aftermath of the most ferocious storm they had ever witnessed, they decided to look for their tent. It should have been impossible; there was almost no chance of finding it. But they went looking anyway. And perhaps the hymns had worked, because – miraculously – there it lay, intact, closed up like a furled umbrella, less than a kilometre away. Now they knew they would live.

Cherry-Garrard described Wilson and Bowers as 'gold, pure, shining, unalloyed. Words cannot express how good their companionship was.'[13] Both of them perished with Scott on the way back from the Pole, leaving Cherry-Garrard haunted by their deaths.

The eggs they collected are now housed by the Natural History Museum in its tiny outpost in the Hertfordshire town of Tring. One of the embryos is there too, sitting on a shelf in a jar of spirits, a forlorn white scrap with bulbous eyes, soft beak and tiny, perfectly formed wings. The remaining two embryos were passed from scientist to scientist, until 1934, when C. W. Parsons of the University of Glasgow finally concluded they had not 'greatly added to our knowledge of penguin embryology'.[14]

Cherry-Garrard was a romantic, especially about the process of discovery. 'Science is a big thing if you can travel a Winter Journey in her cause and not regret it,' he wrote.[15] And though the samples they collected turned out to be scientifically useless, he didn't regret it. Not one bit. With his two companions, he was the first person in the world to see emperor penguins in the wintertime, eggs balanced on their feet to protect them from the sea ice, huddling together against the cold and wind and darkness.

'After indescribable effort and hardship we were witnessing a marvel of the natural world, and we were the first and only men who had ever done so,' he wrote. 'We were turning theories into facts with every observation we made.'[16] He, Wilson and Bowers were the first to share the penguins' world and the first almost to perish in it. And they did it all through sheer bloody-minded, insane, heroic effort.

The next morning, I walked back down on my own for another look at the Adélies. I passed Shackleton's hut, stumbling slightly on the stubbly volcanic rock streaked white with old guano. A skua down in a hollow began beating its wings and scolding me. When it saw it had my attention it took off, flapping in an unnecessarily showy way. Beyond it I could just see the egg on the ground that it was trying to distract me from.

I steered politely away and continued down to the colony, where I found a warm spot out of the wind. I was careful not to get too close to the penguins. The Antarctic Treaty forbids you from approaching any wildlife here – though it's OK if they come up to you. (To do the work that David and all the other researchers do here requires careful scientific justification and vast numbers of forms and permits. Even to visit here I had to be included on one of his permits, but after what he said I had no intention of violating the penguins' personal space, or, indeed, their sense of self.)

My vantage point was surprisingly restful. Most of the penguins were lying on nests, chattering vaguely to themselves. Occasionally, a bird trotted by for what I now recognised as a nest relief. Several were returning to empty nests. Thanks to the blocking sea ice the journey for food had taken too long. Their mates had finally given up and left, the eggs had been stolen by skuas, and most of the stones had been spirited away by other penguins. All that was left was a slight hollow in the rocks and a pitiful few stones that nobody else wanted. But the incoming birds sat there anyway. Occasionally one would stand and stretch its body and neck until it was comically elongated and then snapped back like a rubber band into the normal penguin shape. Then it would flap its flippers madly. This curious ritual passed in waves throughout the colony, as contagious as a yawn.

From where I sat, I could see a lone penguin heading off for food. It looked very thin. Perhaps it was one of the males that had just been relieved. If so, he hadn't wasted much time getting out of there. I watched as he skipped down the slope and hopped over on to the sea ice. He looked like one of the seven dwarfs, on his way to work. Chest thrust out, flippers held out rigidly for balance, he trotted busily along, rocking from side to side, the embodiment of industry and effort. Hi ho, hi ho.

Yesterday evening, over camp dinner, I'd reminded David of Apsley Cherry-Garrard's comment that no creature on Earth had a more miserable existence than an emperor penguin. I'd said that all Antarctic penguins seemed to have a tough life, and I wouldn't like to be reincarnated as any of them.

'It's definitely you against the world if you're an Adélie,' David had replied. 'There's a lot of things conspiring to extinguish your life force. You've got the ice, the ocean, big waves. You're trying to negotiate your way back to your colony, and land on a beach that's being pummelled by ice chunks that weigh tonnes, and there are leopard seals hanging around wanting to eat you. Even back at the colony you still have to worry about stone thieves and skuas, and when, or whether, your mate will come back to relieve you. But I dunno. They seem to smile a lot. They're probably happy.'